Lecture Notes in Biomathematics

Managing Editor: S. Levin

78

Jennifer J. Linderman
Douglas A. Lauffenburger

Receptor/Ligand Sorting Along the Endocytic Pathway

Springer-Verlag

Berlin Heidelberg New York London Paris Tokyo

Authors

Jennifer J. Linderman
Department of Chemical Engineering, University of Michigan
Ann Arbor, MI 48109, USA

Douglas A. Lauffenburger
Department of Chemical Engineering, University of Pennsylvania
Philadelphia, PA 19104, USA

Mathematics Subject Classification (1980): 92

ISBN-13: 978-3-540-50849-6   e-ISBN-13: 978-3-642-48892-4
DOI: 10.1007/978-3-642-48892-4

2146/3140-543210

ABSTRACT

Specialized cell surface macromolecules, termed receptors, are found on nearly all cells and are a means by which a cell can interact with its environment. The receptors are able to bind extracellular ligand molecules and, among other functions, mediate the transport of ligands into the cell itself. A wide variety of macromolecules (e.g. hormones, nutrients, antibodies) are internalized by cells using this highly selective process.

Once inside the cell, receptor/ligand complexes are separated or sorted to appropriate pathways. Typically, receptors are recycled to the cell surface to bind more ligand molecules while ligands are delivered to lysosomes for degradation; however, there can be other outcomes depending on the experimental conditions and systems studied. The problem of how a cell is able to sort receptors from ligands and target these molecules along different pathways is particularly interesting because of the effect of the separation on the overall receptor/ligand processing dynamics. If the cell is able to control the fraction of internalized receptors that are not recycled, it can essentially modulate its response to future doses of the same ligand. An understanding of this modulation can be exploited in the choice of a ligand, for example, a growth factor, to control cell behavior in culture.

The development of an understanding of the basic mechanisms of this separation process is presented, with an emphasis on discovering the fundamental and measurable parameters that influence the process. Mathematical models of the sorting process which include the effects of receptor and ligand diffusion, membrane convective currents, receptor/ligand reaction kinetics, and aggregation of receptors are evaluated. Model predictions are compared to data on several receptor/ligand systems. The model framework developed here may have possible applications to other intracellular sorting processes, for example, the sorting of secreted proteins to the regulated and constitutive pathways.

The authors gratefully acknowledge the support provided by the American Association of University Women (J.J.L.), a National Science Foundation P.Y.I. award (D.A.L.), Monsanto Corporation (D.A.L.), and Merck, Sharp, & Dohme (D.A.L.).

TABLE OF CONTENTS

CHAPTER 1
INTRODUCTION AND OVERVIEW

Eucaryotic cells are able to direct the traffic of many different molecules simultaneously: macromolecules are synthesized and may then be slated for immediate release to the environment, packaged into vesicles for later release to the environment upon the receipt of an appropriate stimulus, incorporated into the cell membrane or into organelle membranes, delivered to lysosomes to function there as enzymes, or delivered to other organelles. Molecules are also brought into the cell from the environment and separated from the receptors which collect and internalize them; these newly internalized molecules can be returned to the environment, routed to lysosomes for degradation, or sent to other intracellular locations. At the same time, their receptors may be returned to cell surface or targeted to lysosomes and other intracellular locations. The ability of the cell to effectively control the traffic of many types of macromolecules, targeting molecules for a variety of destinations based on the type of the molecule and the current needs of the cell, is crucial to cell function.

We focus in this text on one very specific molecular traffic pattern in the cell and the point in that pattern at which the cell chooses the ultimate destinations of the molecules present there. Receptors, specialized macromolecules embedded in a cell's membrane, are responsible for the binding and subsequent internalization of extracellular ligand molecules via the process termed receptor-mediated endocytosis. Once inside the cell, receptors and ligands are separated from each other and directed along different intracellular pathways. Typically, most ligands are delivered to lysosomes and most receptors recycle to the cell surface to bind more ligand molecules. The mechanism of this sorting process is not known. Clearly, the kinetics and efficiency of this sorting step affect the overall kinetics and efficiency of the entire endocytic cycle.

An understanding of the mechanism of intracellular receptor/ligand sorting may lead to the identification of parameters which control the outcome of the sorting process. It is possible that these parameters may be ones that can be controlled or varied, thus enabling the manipulation to some extent of the sorting process and thus of the overall kinetics of the endocytic cycle. For example, it might be advantageous to control a cell's ability to internalize a growth factor or pharmaceutical agent and thereby its response to those molecules.

In this text, the feasibility of several possible sorting mechanisms is investigated. Experimental data on receptor-mediated endocytosis and the sorting step in particular are reviewed in Chapter 2, and the minimal requirements that any proposed sorting mechanism must meet are described. We then investigate equilibrium and simple kinetic sorting schemes and propose in Chapter 4 a mechanism that meets these minimal requirements.

In Chapter 5, we develop a whole cell model for all of the steps of endocytosis. Because experimental data are taken for whole cells, on the overall kinetics of the endocytic cycle and not on the isolated sorting process itself, this is a necessary step in the extraction of detailed sorting data from whole cell data. This model is then used to show that the sorting efficiency may vary for a particular receptor system upon changes in ligand properties. The sorting mechanism proposed in Chapter 4 cannot account fully for these effects. Thus in Chapter 6 we propose and analyze a more complete model for sorting. This mechanism is successful in explaining a variety of outcomes of the sorting process upon changes in several fundamental parameters and provides experimentally testable predictions. Conclusions and recommendations from this study are given in Chapter 7.

CHAPTER 2
RECEPTOR - MEDIATED ENDOCYTOSIS AND
RECEPTOR / LIGAND SORTING

2.1 The endocytic cycle

Receptors are macromolecules that cells embed in their lipid bilayer plasma membranes in order to bind particular extracellular macromolecules, or ligands. Typically glycoproteins, receptors possess an extracellular or ligand-binding domain, a transmembrane domain, and an intracellular or cytoplasmic domain, as shown in Figure 2.1.

The binding of ligands to cell surface receptors can result in the transport of these same ligands to the cell interior or in the adhesion of the cell to a ligand-coated surface. The interaction may also lead to signal transduction, the translation of the binding event into an intracellular sequence of messages and an eventual response. These functions demonstrate that receptors are used by the cell to sense and respond to its environment.

In this text, we focus on the first of these receptor-mediated functions: the internalization of extracellular ligand, a process known as receptor-mediated endocytosis (RME). The purposes of internalization are multiple and dependent on the receptor-ligand system involved. Nutrients too large to diffuse into the cell through membrane pores are endocytosed; included in this category are iron and cholesterol, internalized while bound to the iron-transporting protein transferrin and the cholesterol-transporting low density lipoprotein. Other molecules, such as asialoglycoproteins and immune complexes, are internalized for the purpose of removing them from the extracellular medium. Hormones, growth factors, and other effector molecules (including pharmacologic agents), which may exert some or all of their effects while bound to surface receptors via the signal transduction function of receptors, are internalized. Finally, viruses and toxins also gain entry into the cell via endocytosis. Several ligands, their receptors, and the corresponding reasons for internalization are listed in Table 2.1. Clearly, the process of receptor-mediated endocytosis plays a role in wide variety of cell functions.

The basic steps of receptor-mediated endocytosis or the endocytic cycle have been described in a number of reviews (Brown et al. 1983; Wileman et al. 1985a; Stahl and Schwartz 1986; Helenius et al. 1983; Steinman et al. 1983) and are diagrammed in Figure 2.2. Receptors, of which approximately 10^4 to 10^6 may be present on the cell surface, bind their ligands noncovalently with a high affinity that is characterized by an equilibrium binding

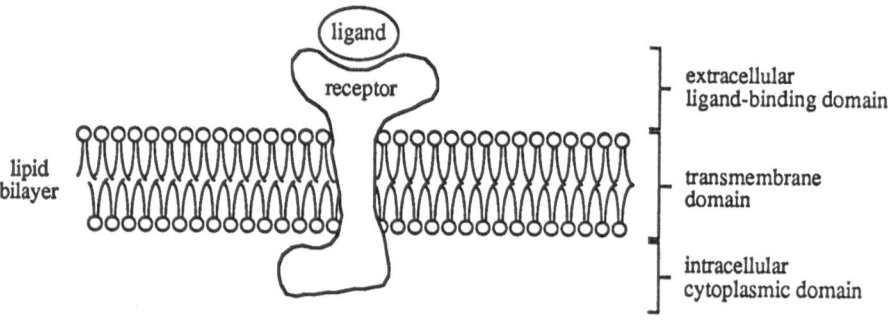

FIGURE 2.1 Receptor embedded in the plasma membrane. Receptors often extend through the lipid bilayer, exposing on one side a ligand-binding domain and on the other side a region that is important for signal transduction.

ligand	receptor	purpose of internalization	reference
low density lipoprotein (LDL)	LDL receptor	nutrition (cholesterol delivery)	Brown and Goldstein 1986
transferrin	transferrin receptor	nutrition (iron delivery)	Klausner et al. 1983
vitellogenin	vitellogenin receptor	nutrition (assembly of yolk body)	Telfer et al. 1982
mannose-6-phosphate glycoproteins	mannose-6-phosphate receptor	retrieval of lysosomal enzymes	Brown et al. 1986
mannose terminal proteins	mannose/N-acetylglucosamine receptor	retrieval of lysosomal enzymes	Stahl et al. 1980
asialoglycoproteins (ASGPs)	ASGP receptor	clearance	Ashwell and Harford 1982
immune complexes	Fcγ receptor	clearance, elicit cellular response	Mellman et al. 1983
epidermal growth factor (EGF)	EGF receptor	elicit cellular response	Cuatrecasas 1982
insulin	insulin receptor	elicit cellular response	Marshall 1985a
chemoattractant	chemoattractant receptor	elicit cellular response	Zigmond et al. 1982
immunoglobulin E (IgE)	IgE receptor	elicit cellular response, clearance of antigen	Furuichi et al. 1985
influenza virus	?	infection	Marsh 1984
diphtheria toxin	?	infection	Olsnes and Pihl 1982

TABLE 2.1 Several receptor-ligand systems which exhibit receptor-mediated endocytosis. Ligand molecules include nutrients, molecules to be cleared from the body, hormones, growth factors, viruses, and toxins. In some cases, responses elicited by ligand binding at the cell surface may be attenuated by internalization; alternatively, internalization may be required for the response.

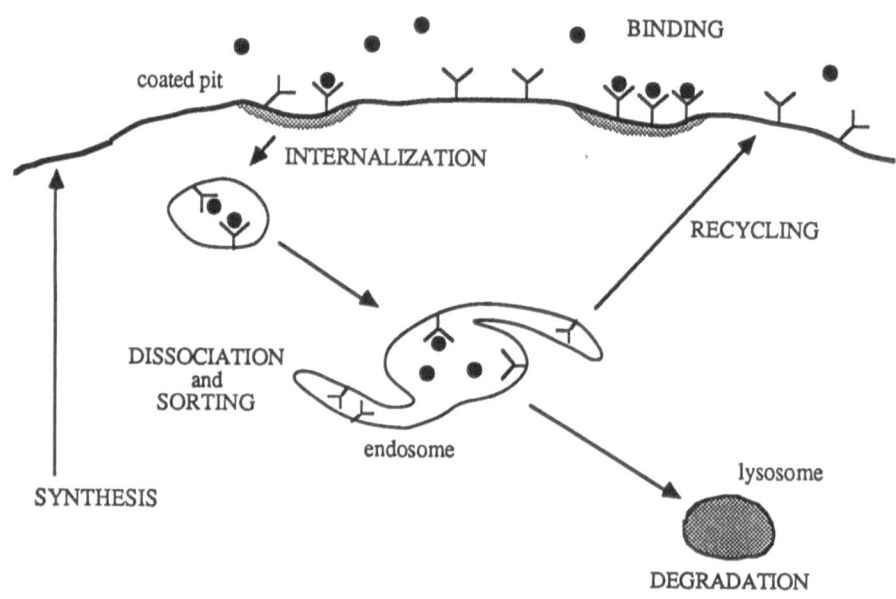

FIGURE 2.2 The endocytic cycle. Receptors mediate the transport of ligand from the extracellular environment to the interior of the cell. Receptor and ligand molecules are sorted intracellularly in the endosome, typically allowing receptor recycling to the cell surface and ligand delivery to lysosomes.

constant on the order of 10^7 to 10^{10} M^{-1}. Physiologic ligands are typically present in very low concentrations, on the order of 10^{-7} M or less, so this high binding affinity is necessary if an appreciable number of ligand molecules are to be bound. The specificity and high affinity of this receptor/ligand binding enable the cell to select exactly the molecules in its complex medium with which it will interact and to which it will respond.

Receptors and receptor/ligand complexes are free to diffuse on the two-dimensional membrane surface. Translational diffusion coefficients on the order of 10^{-10} cm^2/sec have been measured for a variety of receptor systems (Menon et al. 1986a; Schlessinger et al. 1978; Hillman and Schlessinger 1982; Giugni et al. 1987). The most popular technique for measuring the diffusion coefficient of the complex is termed fluorescence photobleaching recovery, or FPR (Edidin et al. 1976). Ligands are labeled with a fluorescent compound such as rhodamine or fluorescein and allowed to bind to surface receptors. A very narrow band of laser light is used to bleach out the fluorescence on a small spot of the cell surface, and the recovery of fluorescence in that spot, the result of the diffusion of nearby receptors with their unbleached ligands into the bleached region, is monitored and used to calculate the diffusion coefficient.

The diffusion of bound receptors is thought to be the primary mechanism in allowing these receptors to reach coated pits, specialized regions of the cell surface about 0.1 µm in diameter that are characterized by the presence of the 180 kD protein clathrin (Goldstein et al. 1984). The bound receptors are trapped in the coated pits. The mechanism of this trapping is not known and may require clathrin or another coated pit-associated protein (Robinson 1987) to function as a "receptor for a receptor", that is, as a protein which specifically binds certain receptors that have diffused into the pits. In some cases it is also known that free receptors may cluster in coated pits prior to ligand binding; this has been shown for the low density lipoprotein receptor (Anderson et al. 1982). By serving as a sink for receptors, the coated pits serve to trap and effectively concentrate receptor/ligand complexes prior to internalization.

The complexes are internalized when coated pits invaginate and pinch off to form small intracellular coated vesicles, about 50 to 150 nm. in diameter, containing the complexes. This process does not occur at temperatures below 10 °C and occurs only slowly at temperatures between 10 and 20 °C (Weigel and Oka 1981). The coated vesicles also contain a small amount of the surrounding medium, internalized nonspecifically together with the specific internalization of any ligands bound to receptors in the coated pits. The coating of clathrin is rapidly removed from the vesicles by an ATP-dependent enzyme, leaving small uncoated vesicles. There is recent evidence that these endocytic vesicles then fuse to form a larger intracellular body, the endosome. To demonstrate this, Braell (1987) developed an assay for the fusion of endocytic vesicles in a cell-free environment. Two sets of Chinese hamster ovary (CHO) cells were allowed to endocytose a ligand, avidin-linked β-galactosidase for the first and biotinylated IgG for the second set of cells. Endocytosis was allowed to continue for about five minutes, in order to assure that the ligands were present primarily in the endocytic vesicles. The cells were next fractionated and the extracts containing containing the endocytic vesicles were isolated. Extracts from group one and group two cells were then mixed and incubated. The formation of an avidin- β-galactosidase- biotinylated IgG complex was measured and indicated that fusion between vesicles had occurred. Similar experiments with comparable results have been performed by other investigators (Davey et al. 1985; Gruenberg and Howell 1986).

The flow cytometry experiments of Murphy (1985) also suggest the fusion of small primary endocytic vesicles. 3T3 fibroblasts were incubated with fluorescein isothiocyanate (FITC)-conjugated dextran at 37 °C. During this time, the dextran was internalized nonspecifically due to its presence in the bulk fluid which surrounds the cells and the endocytic vesicles became loaded with this fluorescent marker. After variable periods of incubation, cells were washed and homogenized to produce cell lysates containing the fluorescent vesicles. Flow cytometric analysis of the lysates revealed kinetically distinct vesicle classes. Most importantly, an early weakly fluorescent population of vesicles gradually decreased in frequency as a more highly fluorescent population emerged, suggesting the fusion of small primary endocytic vesicles with each other in the formation of larger vesicles or endosomes.

Within the endosome, a sorting, or separation, process occurs. The mechanism of this sorting process is not known. Some of the molecules, in most cases primarily receptors, are returned to the cell surface. This is termed receptor recycling or, in the case of a ligand, exocytosis or diacytosis. Receptors which recycle are able to participate in future rounds of endocytosis. For example, the low density lipoprotein (LDL) receptor is known to make about 150 trips through the cell in its 30 hour lifespan (Brown et al. 1983). The ability of receptors to be reused has been demonstrated in many systems by the continuing accumulation of intracellular ligand without a concurrent decrease in available receptors and by the inhibition of uptake by inhibitors of receptor recycling (Schwartz et al. 1982; Schwartz et al. 1984; Ciechanover et al. 1985; Gonzalez-Noriega et al. 1980; Basu et al. 1981; van Leuven et al. 1980; Marshall 1985b; Stahl et al. 1980).

In many cells, the remaining molecules, typically ligands, are sent along a second pathway and delivered to lysosomes. Lysosomes, membrane-bounded vesicles of approximately 0.2 to 0.5 µm in diameter, contain hydrolytic enzymes to digest these molecules. These enzymes perform optimally at a low pH, and therefore lysosomal proton pumps maintain the lysosomal pH at about 4.6 to 5.0. Most molecules delivered to the lysosome, receptors or ligands, are degraded and fragments are released for use by the cell or for disposal in the surrounding medium. It is in the lysosome, for example, that the protein LDL is degraded to release free cholesterol for future use by the cell. At least one ligand is not degraded; mannose-6-phosphate ligands are delivered to the lysosome and function there as lysosomal enzymes.

Receptors are often able to complete the entire endocytic cycle in just a few minutes. For example, the bound transferrin receptor was found to enter HepG2 cells (a human hepatocyte-derived cell line) with a mean time of 3 to 5 min. and return to the surface with a mean time of about 7 min. The bound asialoglycoprotein receptor was found to require a mean time of 2.2 min. for internalization and 4.2 min. for recycling in hepatic parenchymal cells. In both cases, complete cycle times of approximately 15 min. were measured; the ligand binding step for a ligand concentration of 50 nM asialoglycoprotein or 77 nM transferrin was included in this calculation (Ciechanover et al. 1985).

The dynamics of the endocytic cycle are also influenced by the synthesis of new receptors, occurring in most cells at a slow but measurable rate. Mellman and Plutner (1984) depleted J774 cells (a mouse macrophage cell line) of over half of their surface Fc_γ receptors and then measured the subsequent increase in surface receptors with time. Receptors appeared on the surface at the rate of about 5% of their original undepleted number per hour. That this appearance represented newly synthesized receptors was demonstrated by a duplicate experiment in which cycloheximide, an inhibitor of protein synthesis, was included in the medium. In this second experiment, no increase in surface receptor number was observed. Similarly, Marshall (1983) found that only about 2% of the total insulin receptor number in rat adipocytes was synthesized per hour.

It has been demonstrated that in some sytems the rate of new receptor synthesis may vary with time, increasing, for example, when a cell senses a deficiency of a particular ligand. Weissman et al. (1986) incubated K562 cells (a human erythroleukemia cell line) for 6 hours at 37 °C under various media conditions and used ^{35}S-methionine to label the biosynthetic pathway and thus detect new receptor synthesis. Cells provided with a minimal source of iron were found to synthesize transferrin receptors at a greater rate than cells exposed to saturating amounts of diferric transferrin (transferrin carrying bound iron). A still higher rate of synthesis was exhibited by cells that were continually depleted of their receptors by a monoclonal antibody treatment. Similarly, Earp et al. (1986) have demonstrated that epidermal growth factor (EGF) stimulates the synthesis of EGF receptors in WB cells (a nontransformed line of rat liver epithelial cells) in a dose-dependent manner, and thus the receptor degradation normally observed in these cells upon exposure to EGF is offset by this new synthesis.

2.2 The sorting step in the endocytic pathway

The sorting of receptors and their ligands occurs in the endosome and is a critical step in the endocytic pathway, for at this point the cell determines the destinations of the endocytosed molecules. If receptors are efficiently recycled, these receptors can participate in future rounds of endocytosis. On the other hand, by routing receptors to the degradative pathway or to another intracellular location instead of back to the cell surface, the cell can decrease or "downregulate" the number of cell surface receptors and therefore also its ability to respond to future doses of the same ligand. As one example, the mitogenic response of fibroblasts to epidermal growth factor (EGF) may be influenced by the degree of recycling of EGF receptors (Lauffenburger et al. 1987). Clearly, then, an understanding of the sorting mechanism is critical to an understanding of the kinetics and efficiency of receptor recycling and downregulation, of ligand degradation and exocytosis, and of the overall kinetics of the endocytic cycle.

Evidence for this sorting step in different systems has been found by investigators using a variety of biochemical and morphological techniques to show that receptors and ligands are found in the same intracellular compartments immediately after internalization but rapidly move to separate compartments. In three such studies, Evans and Flint (1985), Baenziger and Fiete (1986), and Mueller and Hubbard (1986) found that in rat hepatocytes various asialoglycoprotein ligands pass from a vesicular compartment containing the asialoglycoprotein receptor in appreciable amounts to a compartment that does not.

Endosome morphology and its relation to endosome function

Electron micrographs of endosomes show that they consist of a central vesicular chamber, approximately 0.2 to 0.8 μm in diameter, and one or more attached thin tubules, about 0.01 to 0.06 μm in diameter (rat hepatocytes: Gueze et al. 1983, Geuze et al. 1987, and Wall et al. 1980; BHK-21 cells: Marsh et al. 1986; giant HeLa cells: Bretscher and Thomson 1985; rat macrophages, rat reticulocytes, and human fibroblasts: Harding et al. 1985; Chinese hamster ovary cells: Yamashiro and Maxfield 1984). Marsh et al. (1986) have performed a detailed study on the morphology of endosomes in BHK-21 (hamster kidney) cells. Cells were exposed to horseradish peroxidase as a fluid phase (nonspecifically internalized) marker for 15 min at 37 °C. Thin sections (0.2 to 0.5 μm) of the cells were prepared and examined. The outline of the endosome was traced, and computer-aided reconstruction of serial sections enabled the compilation of a three-dimensional image of the endosome. In these cells, the endosome volume was found to be about 0.04 μm^3/endosome; 60 to 70% of this volume was contained within the vesicle. Tubules accounted for the majority of the surface area, about 60 to 70% of the total surface area of 1.5 μm^2/endosome. As many as seven tubules were attached to a single vesicle. Tubules did not appear to connect two adjacent vesicles, so that each vesicle with its associated tubules is a separate entity, an endosome.

The unusual structure of the endosome is related to the way that it is believed to function.

The tubules are thought to be intermediates in the recycling of molecules back to the cell surface. Thus receptors or ligands found in tubules when they break their connection with the vesicle will be recycled. Molecules remaining behind in the vesicle are assumed to be degraded when the vesicle either delivers its contents to or matures into a lysosome.

The first evidence for this theory came from the work of Geuze and coworkers (1983) who identified the endosome as the sorting chamber for the endocytic pathway. They infused rat livers with an asialoglycoprotein ligand, asialoorosomucoid or asialofetuin, in order to load the endocytic pathway with receptors and ligand. The tissue was then prepared as ultrathin (100 nm.) cyrosections. Receptors were next located using monoclonal antibodies and these antibodies were subsequently tagged with 5 nm. (or 8 nm.) colloidal gold particles. Ligand molecules were located in a similar fashion, using monoclonal antibodies to the ligand and tagging with 8 nm. (or 5 nm.) gold particles. This double-label immunogold electron microscopic technique made it possible to locate both the asialoglycoprotein receptor and its ligand simultaneously and to identify the cellular compartment in which the segregation of receptor and ligand appeared to be occurring. The two types of molecules were found together in early endocytic vesicles; here, the ligand label was positioned close to the limiting membrane of the vesicle, indicating that the ligand was in all likelihood still bound to its receptor. Most significantly, receptors and ligands were found together yet in separate domains of a vesiculotubular structure, indicating that the sorting of the two types of molecules was occurring here. In this organelle, receptors were found along both the vesicle and tubule membranes but in much greater concentration along the tubule membranes. Ligand molecules were found primarily in the lumen of the vesicle. That the ligand molecules were no longer associated with the membrane of the vesicle indicated that the ligands had dissociated from their receptors. Geuze and coworkers named the vesiculotubular structure CURL, an acronym for "compartment of uncoupling of receptor and ligand". This compartment is more generally known today as the endosome.

Geuze and coworkers (1983) also noted that endosomes closest to the sinusoidal membrane of the cells (where internalization occurs) had smaller vesicles than those in the lysosomal area of the cell. As the vesicle size increased, the amount of ligand in the lumen of the vesicle increased and the number of receptors in the vesicle appeared to decrease. These data suggest both that there may be repeated fusion of incoming vesicles containing receptors and ligands with a maturing endosomal structure and that receptors may move from the vesicle into a connecting tubule. Because the receptors in these hepatocytes are known to recycle efficiently to the cell surface and the ligand is known to be degraded in lysosomes, the data are consistent with the theory that the tubules of the endosome become or empty into a vehicle that returns its contents to the cell surface and the vesicle of the endosome delivers its contents to or matures into a lysosome.

Geuze, Slot, and Schwartz (1987) recently obtained more quantitative evidence on the separation of receptors in the endosome. As in the earlier study, colloidal gold was employed to mark the location of receptors in the endosomes of hepatocytes. The number of gold particles per micron of vesicle or tubule membrane was determined in a section thickness of 100 μm. It was found that endosomal vesicles contained a wide range of receptor densities, from 1.7 to 30.7 gold-labeled receptors per micron of membrane and that the highest densities occurred in the smallest vesicles. Endosomal tubules, on the other hand, displayed a relatively constant receptor density of 45.8 (± 3.72) gold-labeled receptors per micron of membrane. Although these numbers cannot be directly compared because of the differences in vesicle and tubule diameter as compared with the section thickness, they do indicate a dramatic difference in receptor density between vesicle and tubule. Because these receptors are known to recycle efficiently and because the data suggest the progressive accumulation of receptors within tubules, the data support the notion that the endosomal tubule contents are recycled to the cell surface in some manner.

Another study also supports the idea that the movement of recycled material is from vesicle to tubule(s). Hatae and coworkers (1986) used horseradish peroxidase (HRP) as a nonspecific tracer of the endocytic pathway in kidney proximal tubule cells. They found that HRP appeared in small endocytic vesicles 0.5 min. after exposure to ligand, in larger endocytic vesicles about 1 min. after exposure, and in apical tubules, which were frequently connected to the labeled endocytic vesicles, in increasing amounts from 3 to 7 min. after exposure. Thus the tubules receive the marker HRP from the endocytic vesicles.

Acidity of the endosome

The previous data suggest that receptors and ligands, once dissociated, distribute to different regions of the same endosome: recycling receptors to tubules and ligands slated for degradation to the lumen of the vesicle. Before this redistribution can occur, however, receptor and ligands must dissociate, a process facilitated by the acidic environment of the endosome.

Tycko and Maxfield (1982) exposed cultured mouse fibroblasts to fluoroscein-α_2-macroglobulin, a ligand with a pH-sensitive fluorescent tag, for 15 to 20 min. at 37 °C. This short incubation time insured that any internalized ligand would be found in prelysosomal compartments along the endocytic pathway. Analysis of individual cells by fluorescence microscopy indicated that the ligand was present in endosomal compartments with a pH of 5.0. Similar measurements, in the range of pH 5.3 to 5.5, have been obtained for endosomes in Hep G2 cells and K562 cells (Tycko et al. 1983, van Renswoude et al. 1982). Yamashiro and Maxfield (1984) measured the pH of CHO endocytic compartments containing α_2-macroglobulin or transferrin. They found that α_2-macroglobulin was present in endosomes of pH 5.0 to 5.5, while transferrin was present in a compartment measuring pH 6.4. Because transferrin in a ligand that is exocytosed rather than degraded, they suggested that the pH 6.4 compartment represented the "postsegregation recycling compartment" which we interpret in this context as detached tubules or the compartment receiving the tubule's contents.

The discovery of the acidic environment of the endosome led to an investigation into the mechanism of that acidification. It appears that ATP-dependent proton pumps exist in the endosomal membrane to lower the pH (Galloway et al. 1983; Mellman et al. 1986); further characterization of the mechanism of endosome acidification may be forthcoming due to the recent development of a method by which to isolate endosomes from other cellular organelles of similar properties (Marsh et al. 1987).

Several investigations have been made into the kinetics of *in vivo* endosome acidification. Murphy and coworkers (1984) and Murphy and Roederer (1986) combined flow cytometry with fluorescence measurements to enable the analysis of many cells. Mouse 3T3 fibroblasts were found to acidify the ligands insulin and epidermal growth factor to an endosome pH slightly less than 6.0 within 10 min. After a total of 30 min., the pH slowly decreased to 5.0; a reasonable interpretation of this later decrease is that it represents the fusion of endosomal vesicles with lysosomes or the maturation of endosomes into lysosomes, presumably occurring after recycling molecules have been removed via tubules.

Kielian et al. (1986) followed the kinetics of acidification in BHK-21 cells by using the Semliki Forest virus (SFV), which, like other viruses, fuses rapidly with endosomal membranes at low pH. Wild type and mutant SFV viruses, which fuse with endosomal membranes at a pH of 6.2 and below or 5.3 and below, respectively, were internalized by the cells. Approximately 5 minutes after internalization, the fusion of wild type viruses was detected, indicating that a pH less than or equal to 6.2 had been reached. Some mutant viruses fused immediately after endocytosis and others after several hours, suggesting that the further acidification of the intracellular compartment containing the virus was heterogeneous.

The above summary of experimental data indicates that although a precise determination of endosome pH and the kinetics of acidification has not been made and may in fact differ between cell types and even within a single cell, the endosome pH is likely in the range of 5.0 to 6.0 and the initial acidification of the receptor/ligand complexes probably occurs rapidly after internalization.

The acidification of newly internalized receptor/ligand complexes upon delivery to endosomes promotes receptor/ligand dissociation in many systems. For example, the binding of mannose-6-phosphate ligands, mannosylated ligands, epidermal growth factor, asialoglycoproteins, and low density lipoproteins to their receptors have all been shown to be affected by low pH (Gonzalez-Noriega et al. 1980; Tietze et al. 1982; Wileman et al. 1985b;

Dunn and Hubbard 1984; Hudgin et al. 1974; Basu et al. 1978). The molecular basis for the changes in binding affinity which occur upon acidification are not known, but recent data have shown that the epidermal growth factor and asialoglycoprotein receptors in A431 cells (a human epidermoid carcinoma cell line) undergo conformational changes when the pH is reduced from 7.4 to 5.6 (DiPaola and Maxfield 1984). Similarly, the mannose-6-phosphate receptor shows a change in conformation upon a pH change from 7.5 to 5.4 (Westcott et al. 1987). It seems likely that these types of conformational changes in the receptor, and perhaps also in the ligand, enhance the dissociation constant of the receptor/ligand binding reaction when the pH is lowered. Davis and coworkers (1987) have demonstrated that a particular amino acid sequence contained in many receptors may be responsible for the sensitivity of ligand binding to pH, presumably by undergoing a conformation change.

Some receptor/ligand systems are not affected by the low pH of the endosome, and this is reflected in the sorting outcome. Transferrin is one such ligand. Diferric transferrin binds tightly to cell surface receptors, and it remains bound to its receptor inside the endosome, although the low pH does effect one change: the iron molecules bound by the carrier ligand transferrin are released for use by the cell. The apotransferrin molecule (transferrin with no bound iron) then recycles back to the cell surface along with its receptor and is released here into the medium because the apotransferrin molecule, unlike diferric transferrin, has little affinity for its receptor at normal extracellular pH. Thus the sorting process is dramatically altered when the receptor/ligand binding is not affected by low pH; the receptor and ligand are not targeted for different pathways.

The concept of receptor/ligand dissociation followed by receptor and ligand segregation into different intracellular compartments is supported by the data of Wolkoff and co-workers (Wolkoff et al. 1984). Wolkoff assayed for the ability of receptors to rebind their dissociated ligands upon raising the endosome pH. Rat hepatocytes were allowed to endocytose asialoorosomucoid bound to its receptor, the asialoglycoprotein receptor. Inside the cell, these complexes dissociate at the low endosome pH (Bridges et al. 1982). The proton ionophore monensin was used to raise the endosome pH back to extracellular levels and thus allow rebinding of ligands to receptors, an event which could only occur if the two types of molecules were still present in the same intracellular compartment. The ability of the ligand to reassociate with its receptor was found to decrease with increasing time within the cell, data which are consistent with the concept of dissociation and subsequent segregation during the sorting process.

Finally, it is well established that the pH of the endosome, as well as that of the lysosome, can be raised to extracellular levels using acidotropic agents or carboxylic ionophores such as chloroquine, methylamine, monensin, NH_4Cl, and Tris (hydroxymethyl-aminomethane). These agents have been shown not only to drastically reduce ligand degradation in the lysosome, due to the effects of these drugs on lysosomal pH, but also to disrupt the recycling of receptors to the surface, again illustrating the importance of the acidic environment of the endosome (Mellman et al. 1986; Basu et al. 1981; van Leuven et al. 1980; Marshall 1985b; Gonzalez-Noriega et al. 1980; Schwartz et al. 1984; Harford et al. 1983).

Sorting time and the life history of the endosome

There is much evidence to indicate that molecules in the tubules of the endosome are recycled when that tubule breaks its connection with the central vesicle or otherwise delivers its contents to a recycling vehicle and that the molecules not removed from the central vesicle are not recycled and most often degraded. The time that a receptor or ligand is allotted to redistribute to its appropriate compartment, vesicle or tubule, we define as the sorting time. Although exact measurements of the sorting time have not yet been obtained, several related measurements do give estimates or bounds on that time. Clearly, the sorting time is less than the time required for a bound cell surface receptor to internalize ligand and then recycle back to the cell surface. This time has been estimated to be 6 to 7 min. for the asialoglycoprotein receptor in hepatocytes (Schwartz et al. 1982), and 6 min. for the insulin receptor in adipocytes (Marshall 1985b). Evans and Flint (1985) have obtained a more direct estimate of the sorting time in hepatocytes by isolating vesicles containing asialoglycoprotein and insulin receptors and their ligands at various times after internalization. One to two minutes after uptake, receptors and ligands were found together in endosomes or pre-endosomes. By 10 to 15 min. after internalization, the receptors and ligands were no longer found in the same intracellular compartment, indicating that sorting had been completed. Thus it is reasonable to estimate that in many cells the sorting time is on the order of 5 to 10 min.

There are many questions about the endosome that remain unanswered. It is not yet clear, for example, how endosomes, presumably formed through the fusion of endocytic vesicles, gain tubules. Do these tubules represent small vesicular bodies which have fused onto the central vesicle of the endosome for a short time during their passage toward the cell membrane? Or do tubules grow from the central vesicle itself, the membrane needed to form tubules being donated from incoming new endocytic vesicles fusing with the central vesicle? It seems likely that the low pH of the endosome and elements of the cytoskeleton may play a role in the formation of these tubules, but no such role has yet been documented.

In addition, there are at least two models, proposed by Helenius and coworkers (1983), for the "life history" of the endosome. The maturation model asserts that an individual endosome is a transient structure, forming from the fusion of incoming endocytic vesicles and ultimately, after sorting occurs, fusing with a lysosome or slowly transforming into a lysosome via the delivery of lysosomal enzymes. The second model, the vesicle shuttle model, suggests that endosomes are stable structures which continually receive incoming receptor/ligand complexes, sort these molecules, and target each molecule for recycling, degradation, or another fate. Vesicles would therefore be required to shuttle to and from the endosome, delivering unsorted molecules and removing sorted molecules. Recent data of Roederer and coworkers (Roederer et al. 1987) on the kinetics of exposure of ligand to a low pH compartment and to proteolytic enzymes suggest that the maturation model may be correct.

2.3 Estimation of the sorting efficiency

From the available data on the sorting outcome in several systems, it is possible to obtain a rough estimate of the efficiency with which the sorting process must operate. The fraction of the initial number of receptors remaining after cycle n, $F^{(n)}$, can be estimated by

$$F^{(n)} = (1-f_R) F^{(n-1)} + k\theta \qquad\qquad n = 1, 2, 3 \ldots \qquad\qquad (2.1)$$

where f_R is the fraction of receptors not recycled from the sorting process, k is the synthesis rate of new receptors (fraction of initial receptor number synthesized per unit time), θ is the cycle time (time required by a receptor to traverse the entire endocytic cycle), and n is the number of cycles completed. $F^{(0)}$ is equal to one. Marshall (1985b) has shown that rat hepatocytes are able to internalize insulin for four hours with little or no loss in the total number of insulin receptors. During this time, receptors cycle through the cell and thus through the sorting process approximately every 8 min., and new receptors are synthesized at the rate of about 2% of the original number every hour. Thus the loss of receptors per cycle, f_R, can be shown to be equal to 0.003.

Another estimate of the receptor recycling efficiency can be made using the data of Weissman et al. (1986). Prior to the start of the experiment, transferrin receptors on K562 cells were labelled metabolically for two hours using ^{35}S-methionine. The cells were then transferred to normal growth medium and the fraction of radiolabeled receptors remaining was measured as a function of time. The receptor half-life was found to be about 8 hours, and the cycle time was computed to be 12.5 min. Because no synthesis of labeled receptors occurs during the experiment, the synthesis rate k is equal to zero. One can therefore compute that the loss of receptors per cycle, f_R, is approximately 0.018.

These values of the fractional loss of receptors per cycle, $f_R = 0.003$ and 0.018, are only rough estimates of the outcome of the sorting process due to uncertainties in the biological data and expected variability between individual cells. In addition, different cells may choose to operate with different sorting efficiencies according to their need to internalize ligands. In light of these estimates, however, it seems reasonable to expect that any proposed sorting mechanism must be able to account for a return of 95% of the internalized receptors to the cell surface (i.e., $f_R \leq 0.05$). The mechanism must be able to target these intracellular receptors to the recycling pathway within the estimated sorting time of 5 to 10 minutes.

We also note that are instances in which the sorting outcome has been shown to vary with experimental conditions and may fall far short of this maximum efficiency. These systems will be discussed in detail in Chapters 5 and 6, after the presentation and analysis of a sorting mechanism which will account for the minimal requirements presented here.

CHAPTER 3
AN EQUILIBRIUM MODEL OF THE SORTING PROCESS

3.1 Introduction

Receptor and ligand molecules are separated in the endosome on their route through the endocytic cycle. As noted in Chapter 2, 95% or greater of the receptors may be recycled to the cell surface, while much of the ligand is routed to lysosomes. To explain the segregation of the receptor and ligand populations, we first consider a simple equilibrium mechanism based on the geometry of the endosome.

Receptors and ligands are assumed to dissociate rapidly at the low pH of the endosome, as suggested by a variety of experimental data (Gonzalez-Noriega et al. 1980; Tietze et al. 1982; Wileman et al. 1985b; Dunn and Hubbard 1984; Hudgin et al. 1974; Basu et al. 1978). Thus both types of molecules can be expected to diffuse separately within the endosome, receptors diffusing on the two dimensional membrane surface of the endosome much as they do on the cell surface itself and ligands diffusing within the three dimensional lumen of the endosome. We suggest that the endosome tubules, with a higher surface area to volume ratio than the vesicle, accomplish the segregation by an equilibration of receptor and ligand molecules between the two compartments. At equilibrium, the high surface area tubules can be expected to contain most of the receptors, the high volume vesicle, most of the ligand.

The fraction of receptors and ligands in the tubules of the endosome at equilibrium, the fraction that is recycled, can thus be calculated.

3.2 Predicted separation

Marsh et al. (1987) have estimated that 60 to 70% of the volume and 30 to 40% of the surface area of the endosome are contained in its vesicular portion. If this represents a typical endosome geometry, one can predict that 60 to 70% of internalized receptors be found in tubules at equilibrium and will thus be recycled from the endosome. The remaining 30 to 40% of the receptors remain in the vesicle and will ultimately be degraded in lysosomes. This degree of receptor loss contradicts experimental measurements on several systems (Weissman et al. 1986; Marshall 1985).

Because the fraction of receptors recycled in a single stage equilibrium process falls short of that required to account for the experimental data, we also considered multiple stage separations. This type of separation might occur if, for example, a first set of tubules received their equilibrium content of receptors and ligands from the vesicle and then returned to the cell surface while a second set of tubules fused with the central vesicle and the process continued. The results of two and three stage separations are shown in Figure 3.1.

In addition, the fraction of ligand recycled from the endosome and thus exocytosed can be calculated from the fractional volume of the tubules, the number of equilibrium stages, and the partition coefficient κ, defined as

$$\kappa = (1-\lambda)^2, \tag{3.1}$$

where λ is equal to the diameter of the ligand divided by the diameter of the tubule (Pappenheimer 1953). The partition coefficient is equal to the ratio of the concentration of ligand in the tubule at equilibrium divided by the concentration of ligand in the vesicle at equilibrium. The fraction of ligand recycled is shown in Figure 3.1 for one, two, and three stage separations.

3.3   Discussion

For the simple sorting mechanism proposed here, the equilibration of the contents of the endosome according to endosome geometry, the fraction of receptors and ligands recycled is easily calculated. A single stage mechanism does not allow for the efficient receptor recycling that has been experimentally observed, although multiple stages do predict the necessary efficient receptor recycling. However, the fraction of ligand recycled also increases rapidly with the number of separation stages, resulting in a less efficient separation of receptors and ligands as the number of stages increases. Although there are systems in which a high degree of ligand exocytosis is observed, on the order of 50% (Mellman et al. 1984; Townsend et al. 1984), it is the exception and not the usual outcome of sorting.

In addition, there is no experimental evidence suggesting a multistage separation in the endosome; such an operation would require complicated timing schemes and tubule traffic. For example, tubules might be required to fuse with the vesicle, receive receptors and ligands, and break their connection with the vesicle three consecutive times within the sorting time, in each case maintaining the tubule-vesicle connection long enough for an equilibrium distribution of molecules to be approached. It is not at all clear whether the cell could coordinate such a scheme for one endosome, much less for the many endosomes that are present within the cell at one time. Finally, this equilibrium calculation does not indicate the length of time required for

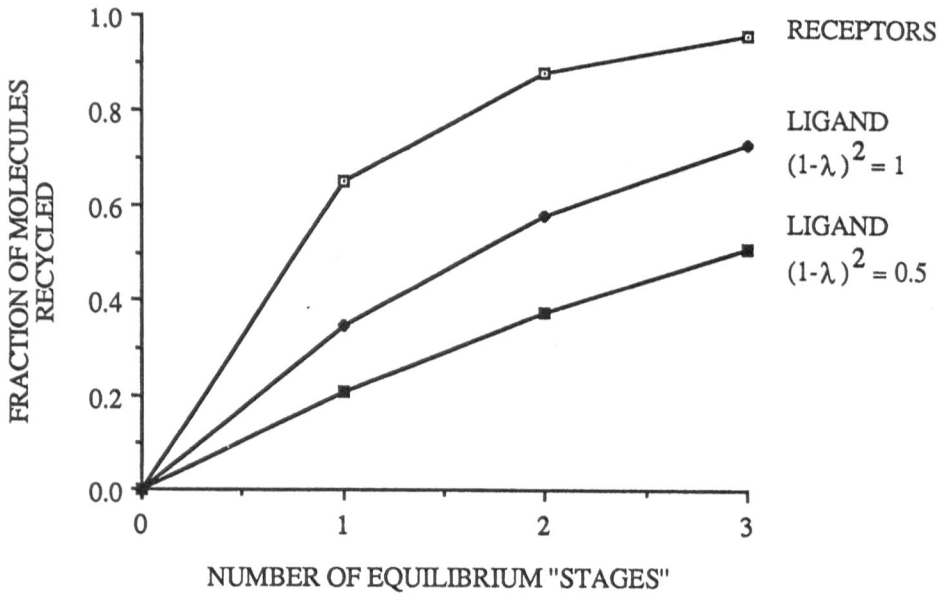

FIGURE 3.1 Predicted separation for an equilibrium sorting mechanism. Plot shows the fraction of receptor or ligand molecules in tubules, and therefore recycled, for one, two, and three stage separations. Points shown were calculated assuming that endosome tubules contain 65% of the surface area and 35% of the volume of the entire endosome.

the receptor and ligand contents of the endosome to reach an equilibrium distribution. The sorting time is estimated at 5 to 10 minutes, and at this point in our modeling it is not yet known whether this is sufficient to allow equilibration in one or multiple stages.

For these reasons, we feel that although the geometry of the endosome is helpful to its sorting function, the equilibrium mechanism is an unlikely candidate for the sorting mechanism. In the next chapter, we consider two possible kinetic schemes for intracellular receptor/ligand sorting based on the diffusion of the ligand and receptor molecules in the endosome.

CHAPTER 4
SIMPLE KINETIC MODELS OF SORTING

4.1 Introduction

In Chapter 3, the distribution of receptors and ligands in the endosome at equilibrium was examined and found to give a poorer separation than has been experimentally observed in several systems. In addition, this scheme does not allow for a variety of outcomes of the sorting process, as suggested by experimental data discussed later in this text (Chapters 5 and 6). Furthermore, it is not known whether the kinetics of ligand and receptor movement into tubules are rapid enough to allow an equilibrium distribution of molecules to be approached within the estimated sorting time. For these reasons, models of sorting which include the kinetics of receptor and ligand transport between the vesicle and tubules should next be examined.

In an attempt to elucidate the kinetic mechanism by which intracellular receptor/ligand sorting is accomplished, we make several assumptions. The beginning of the sorting process is defined as the time that internalized receptors and ligands are first located in the endosome, a low pH environment with the geometry of a central vesicle and attached thin tubules. All receptors and ligands are assumed to be initially located in the vesicle, presumably the result of the fusion of several smaller vesicles formed at the cell surface (Braell 1987; Davey et al. 1985; Murphy 1985). The sorting process ends with the detachment of the tubule(s) from the central vesicle at a time we define as the sorting time. Any receptors and ligands in tubules at the end of this sorting time are assumed to be recycled, and molecules remaining behind in the vesicle are assumed to be eventually degraded.

As suggested by the data of Mellman et al. (1984), Dunn and Hubbard (1984), Gonzalez-Noriega et al. (1980), Tietze et al. (1982), Wileman et al. (1985b), Hudgin et al. (1974), and Basu et al. (1978) for the Fc receptor, the epidermal growth factor, the mannose-6-phosphate receptor, the mannose receptor, the asialoglycoprotein receptor, and the low density lipoprotein receptor systems, which show that the receptor/ligand binding is sensitive to the low pH of the endosome, the dissociation of the receptor and ligand complex will be assumed to be rapid so that, for the purposes of these simple kinetic models of sorting, the two types of molecules can be treated separately. In particular, we are interested in the separate movements of receptors and ligands from the vesicle into a tubule because these are the molecules that are recycled. Receptor molecules are assumed to diffuse within the plane of the endosome membrane, as they do on the cell surface (Schlessinger et al. 1978; Wiegel 1984). Ligands are assumed to diffuse within and between the tubule and vesicle volumes.

The movement of these molecules, at least initially, is from vesicle to tubule. This assumption is supported by the observation of Geuze et al. (1983, 1987) who found that the endosome vesicles close to the cell surface, presumably those of the early endosomes, contained many receptors, but that vesicles further into the cell, those of the late endosomes, contained fewer receptors. In addition, Hatae et al. (1986) found that in kidney proximal tubule cells the movement of endocytosed material that is recycled is from vesicle to tubule.

The simple kinetic models of sorting that will be examined require knowledge of the rates at which receptor and ligand molecules enter tubules from endocytic vesicles. Several possibilities for the sorting mechanisms can be considered and examined knowing these transport rates.

We first consider the possibility that receptors are able to segregate from ligand molecules by moving into endosome tubules at a greater rate. The rate constants for the movement of receptor and ligand molecules into tubules are likely to be very different. Receptors, which diffuse on the surface of the vesicle, need only search in two dimensions for a tubule entrance. The ligand, on the other hand, must search in three dimensions. Adam and Delbrück (1968) found that the mean time for a molecule to diffuse to a target is strongly dependent on the dimensionality of the system. While the ligand has the advantage of a diffusion coefficient that is likely two to three orders of magnitude greater than that of the receptor, the receptor experiences the advantage of lower dimensionality in its search. The time it takes each to find a tubule entrance will depend on the vesicle radius, the appropriate diffusion coefficient, the number of tubules connecting with the vesicle, and the size of a tubule entrance.

Using the calculated transport rate constants and appropriate parameter values, one can determine whether the receptors are able to reach the tubule entrance more rapidly than the ligand molecules. The receptor's advantage in dimensionality may result in a greater rate of transport of receptors to the tubule entrance, and we are interested in the circumstances under which this possibility could be realized. Our analysis allows this possible sorting mechanism to be evaluated.

We can also consider the sorting efficiency of the above mechanism when a convective membrane current is present and enhances the rate of movement of receptors into tubules. If tubules bud from vesicles, drawing membrane from the vesicle itself, then it is not difficult to envision a convective membrane current pulling the receptors toward the tubule entrance. This possibility is diagrammed in Figure 4.1. Depending on the magnitude of the convective current, this could be the primary mechanism for moving the receptor population. Therefore, we will also calculate the rate of transport of receptors into a tubule in the presence of a

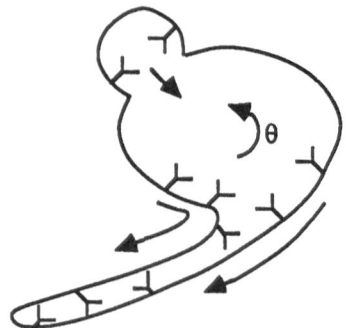

FIGURE 4.1 Convective membrane current. Lengthening tubule draws membrane from vesicle, pulling receptors into the tubule. Membrane might be donated from new endocytic vesicles fusing with the central vesicle of the endosome.

membrane current, assuming that the radius of the vesicle remains constant. A constant vesicle radius is possible if new membrane is donated from vesicles fusing with the central endosome vesicle at the same rate that membrane is removed into the tubule. Thus the analysis will allow the contribution of a convective membrane current to be evaluated.

Second, another possible sorting mechanism is considered. There may be a trapping mechanism in the tubules that prohibits the movement of receptors from the tubule back into the vesicle. This trapping mechanism may be the interaction of receptors in tubules with other molecules, such as clathrin, a similar trapping molecule, or elements of the cytoskeleton, serving to trap and effectively concentrate receptors prior to recycling. Clathrin, a protein thought to play a role in trapping receptors in coated pits on the cell surface, has been noted to be present around the endosome tubules in hepatic parenchymal cells, although it is not yet clear if this clathrin has been associated with the tubule membrane (Geuze et al. 1983; Geuze, H.J., and J.E.A.M. Zijderhand-Bleekemolen, personal communication). More recently, Brown et al. (1986) have found that the endosomes of Clone 9 hepatocytes contain tubular extensions, some of which have coated pits, and they suggest that mannose-6-phosphate receptors are collected and concentrated in the coated pits and/or tubules for recycling. Alternatively, the trapping mechanism may be the existence of a convective membrane current flowing from the vesicle into the tubule, serving to form the tubule as well as to carry receptors into the tubule and trap them there (Figure 4.1). Unlike the receptors, the ligand molecules may diffuse into and out of the tubules and eventually equilibrate between the vesicular and tubular volumes of the endosome. For this mechanism, then, the time required for a significant fraction of the receptors to move into a tubule must be calculated, as well as the amount of ligand that would also be found in the tubules. Thus the second kinetic sorting mechanism we will evaluate is the separation of receptors and ligands in the presence of a receptor trap in the tubules.

4.2 Mathematical models

We want to investigate the movement of receptor and ligand molecules toward a tubule entrance on the surface of the cell's sorting chamber, the endosome. More specifically, the mean time or its inverse, a rate constant, for a molecule of each species to reach a tubule is to be calculated. As shown in Figure 4.2, the endosome is modeled as a sphere of radius R to which is attached a single thin tubule of radius b. The critical angle θ_c is defined as the ratio b/R. We assume that the radii R and b do not change significantly during the sorting time. The assumption that at any given time only one tubule entrance is available to receptors and ligands is reasonable in view of the electron micrographs of Geuze et al. (1983) and is made for simplicity. Calculations might also be done for the case of multiple tubules, as suggested by

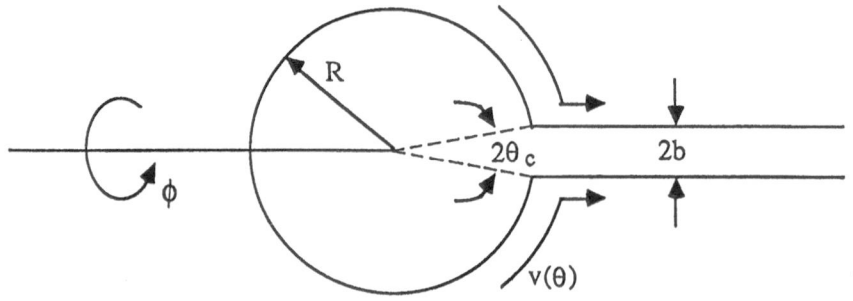

FIGURE 4.2 Model geometry. The endosome is modeled as a sphere of radius R to which is attached a single thin tubule of radius b. The angle θ_c is equal to b/R and defines the size of the tubule opening. There may be a convective membrane current $v(\theta)$ acting in the direction shown to aid in the transport of receptors from the vesicle into the tubule.

the measurements of Marsh et al. (1986), but the general character of the rate constants is likely to be found from the single tubule model.

Initially, receptors are assumed uniformly distributed on the spherical surface and ligand molecules uniformly distributed within the sphere. We temporarily assume that the tubule entrance is perfectly absorbing for receptor and ligand molecules in order to calculate and compare the rate constants for the transport step from the vesicle to the tubule.

Ligand Motion

In related problems, investigators have calculated rate constants for the binding of extracellular ligand to receptors on the surface of a cell (Berg and Purcell, 1977; DeLisi and Wiegel, 1981; Brunn, 1981). The most common approach is to postulate a constant ligand concentration far from the cell and solve the steady state diffusion equation for the region exterior to a sphere with mixed boundary conditions at the sphere surface. For the endosome problem, however, because the ligand molecules in a endosome diffuse within the interior and not on the exterior of a sphere, finding the mean time to reach the tubule entrance would require the solution of an unsteady state diffusion equation. We chose instead a simpler technique.

Berg and Purcell (1977) and Szabo et al. (1980) derive an equation for the mean time required by a molecule moving in n dimensions to reach a target when there is no interaction potential between the molecule and the sink,

$$D_n \nabla_n^2 W + 1 = 0, \tag{4.1}$$

where D_n is the appropriate translational diffusion coefficient and ∇_n^2 is the Laplacian operator in the relevant dimension and coordinate system. The mean time to capture, W, is a function of the initial or starting position of the molecule and is approximately the inverse of the rate constant for a molecule finding the tubule entrance (Szabo et al. 1980). Because a finite mean capture time is generated by the initial placement of a molecule in the vesicle and all initial positions are equally likely, there is a constant source term in the equation.

The mean capture time equation, Eq. 4.1, can be used for the motion of the ligand in the endosome. The equation for the ligand capture time W_L is therefore

$$D_L \left[\frac{1}{r^2} \frac{\partial}{\partial r} \left(r^2 \frac{\partial W_L}{\partial r} \right) + \frac{1}{r^2 \sin \theta} \frac{\partial}{\partial \theta} \left(\sin \theta \frac{\partial W_L}{\partial \theta} \right) \right.$$

$$\left. + \frac{1}{r^2 \sin^2 \theta} \frac{\partial^2 W_L}{\partial \phi^2} \right] + 1 = 0, \tag{4.2}$$

where D_L is the diffusion coefficient for the ligand in the vesicle lumen. Eq. 4.2 is subject to the following boundary conditions:

$$W_L = 0 \qquad\qquad r = R, \ \theta \le \theta_c \tag{4.3}$$

$$\frac{\partial W_L}{\partial r} = 0 \qquad\qquad r = R, \ \theta > \theta_c \tag{4.4}$$

$$W_L (\phi = 0) = W_L (\phi = 2\pi) \tag{4.5}$$

$$\frac{\partial W_L}{\partial \phi} (\phi = 0) = \frac{\partial W_L}{\partial \phi} (\phi = 2\pi) \tag{4.6}$$

$$\frac{\partial W_L}{\partial \theta} = 0 \qquad\qquad \theta = \pi. \tag{4.7}$$

Eqs. 4.5 and 4.6 are periodic boundary conditions and Eq. 4.7 states the condition of boundedness on θ. We assume that the tubule entrance is equivalent to a spherical cap and not the disk formed by removing that cap. For small values of the angle θ_c, the difference in capture times given either of the two alternatives should be negligible.

The mean capture time equation for the ligand is difficult to solve because the boundary conditions on the surface of the sphere are mixed; over part of the surface the boundary condition is that of no flux while over the remainder of the surface an absorbing boundary is specified. This equation is solved using an approximate Green's function technique suggested by the work of Brunn (1981).

To solve Poisson's equation inside a sphere with mixed boundary conditions on the surface, Eq. 4.2, the Green's function for a simpler problem, Laplace's equation with constant flux boundary condition over the entire surface of the sphere, is found. This Green's function is then used to treat both the inhomogeneity in the equation itself and the inhomogeneity in the boundary condition. The boundary condition inhomogeneity can be more easily seen by rewriting boundary conditions in Eqs. 4.3 and 4.4 as

$$\frac{\partial W_L}{\partial r} = 0 \qquad r = R, \; \theta > \theta_c \qquad (4.8)$$

$$\frac{\partial W_L}{\partial r} \neq 0 \qquad r = R, \; \theta \leq \theta_c. \qquad (4.9)$$

The flux of ligand into the sink ($\theta \leq \theta_c$) is an unknown function of the angular position θ. The appropriate Green's function for Laplace's equation is

$$G\,(\,r,\theta,\phi\,;\,r',\theta',\phi'\,) \; = \; \frac{1}{4\pi} \left[\frac{R}{(R^4 + r^2 r'^2 - 2rr'R^2 \cos\gamma\,)^{1/2}} \right.$$

$$+ \frac{1}{R} \ln \left(\frac{2R^2}{R^2 - rr'\cos\gamma + (R^4 + r^2 r'^2 - 2rr'R^2 \cos\gamma\,)^{1/2}} \right)$$

$$+ \; (\,r^2 + r'^2 - 2rr'\cos\gamma\,)^{-1/2} \left. \right], \qquad (4.10)$$

where (r,θ,ϕ) are the coordinates of the observation point r and (r',θ',ϕ') are the coordinates of the source point r'. The angle γ between (θ,ϕ) and (θ',ϕ') is defined by

$$\cos\gamma \; = \; \cos\theta \cos\theta' \; + \; \sin\theta \sin\theta' \cos(\phi - \phi'). \qquad (4.11)$$

The details of the derivation of this Green's function can be found in Appendix I. The Green's function satisfies boundary conditions given in Eqs. 4.5, 4.6, and 4.7 and the constant flux boundary condition

$$\frac{\partial G}{\partial r} \; = \; \frac{-1}{4\pi R^2} \qquad r = R. \qquad (4.12)$$

By defining dimensionless variables for radial position,

$$\xi \; = \; \frac{r}{R} \qquad (4.13)$$

$$\xi' \; = \; \frac{r'}{R}, \qquad (4.14)$$

the Green's function can be rewritten as

$$G(\xi,\theta,\phi\,;\xi',\theta',\phi') = \frac{1}{4\pi R}\left[\,(1+\xi^2\xi'^2-2\xi\xi'\cos\gamma)^{-1/2}\right.$$

$$+\ln\left(\frac{2}{(1-\xi\xi'\cos\gamma)+(1+\xi^2\xi'^2-2\xi\xi'\cos\gamma)^{1/2}}\right)$$

$$\left.+(\xi^2+\xi'^2-2\xi\xi'\cos\gamma)^{-1/2}\right], \tag{4.15}$$

which has dimensions of length^{-1}.

The inhomogeneity in Eq. 4.2 due to the source term $-1/D_L$ is next accounted for by a volume integral, and the inhomogeneous boundary condition (Eq. 4.9) by a surface integral. The general solution to Eq. 4.2 is thus

$$W_L(\xi,\theta,\phi) = \frac{1}{D_L}\iiint_V G(\xi,\theta,\phi\,;\xi',\theta',\phi')\,dV'$$

$$+\iint_{sink}\frac{\partial W_L}{\partial \xi}\bigg|_{\xi=1} G(\xi,\theta,\phi\,;1',\theta',\phi')\,dS' + C. \tag{4.16}$$

The surface integration here and in later integrals is performed only over the surface of the sink because $\partial W_L/\partial\xi$ evaluated at $\xi=1$ is equal to zero elsewhere. The specification of Neumann boundary conditions in finding the Green's function means that the solution for the mean time W_L can only be determined, at this point in the development of the solution, to within a constant C.

Two unknowns in Eq. 4.16 prevent solution for the mean time directly. The first is the radial derivative of W_L appearing inside the surface integral, and the second is the constant C. We now introduce the approximation that $\partial W_L/\partial\xi$ does not vary significantly over the sink and replace the derivative by its average,

$$j = avg\left[\frac{\partial W_L}{\partial \xi}\text{ at }\xi=1,\ 0\le\theta_c\right]. \tag{4.17}$$

This approximation should work well for small values of θ_c and allows the removal of the derivative from within the surface integral. The value of j can be found by applying Fredholm's Alternative (Ramkrishna and Amundson 1985), the requirement of solvability, to the approximate equation

$$
W_L(\xi,\theta,\phi) \cong \frac{1}{D_L} \iiint_V G(\xi,\theta,\phi;\xi',\theta',\phi')\,dV'
$$

$$
+ j \iint_{sink} G(\xi,\theta,\phi;1,\theta',\phi')\,dS' + C \qquad (4.18)
$$

by setting

$$
-\frac{1}{D_L} \iiint_V dV = j \iint_{sink} dS \qquad (4.19)
$$

and solving for j. The result is

$$
j = \frac{-2R}{3D_L(1-\cos\theta_c)}. \qquad (4.20)
$$

To find the remaining unknown, C, in Eq. 4.16, the boundary condition in Eq. 4.3 is applied to require that the mean capture time is zero at the sink, or

$$
0 = \frac{1}{D_L} \iiint_V G(1,\theta^*,\phi^*;\xi',\theta',\phi')\,dV'
$$

$$
+ j \iint_{sink} G(1,\theta^*,\phi^*;1,\theta',\phi')\,dS' + C, \qquad (4.21)
$$

where $(1, \theta^*, \phi^*)$ is a point on the sink. The two integrals are approximated by their average value over the sink, and the dimensionless integrals

$$
\beta_1 = \frac{1}{R^2} \operatorname{avg}\left[\iiint_V G(1,\theta^*,\phi^*;\xi',\theta',\phi')\,dV'\right] \qquad (4.22)
$$

$$\beta_2 = \frac{1}{R} \text{ avg } \left[\iint_{\text{sink}} G(1,\theta^*,\phi^* ; 1,\theta',\phi') \, dS' \right] \qquad (4.23)$$

are defined. The solution for C is then

$$C = \frac{R^2}{D_L} \left[\frac{2\beta_2}{3(1-\cos\theta_c)} - \beta_1 \right]. \qquad (4.24)$$

Combining Eqs. 4.18, 4.20, and 4.24, the mean capture time for the ligand is obtained as

$$W_L(\xi,\theta,\phi) = \frac{R^2}{D_L} F_L(\theta_c; \xi,\theta,\phi), \qquad (4.25)$$

where

$$F_L(\theta_c; \xi,\theta,\phi) = \alpha_1 - \beta_1 + \frac{2}{3(1-\cos\theta_c)}(\beta_2 - \alpha_2) \qquad (4.26)$$

and the dimensionless integrals α_1 and α_2 are defined by

$$\alpha_1 = \frac{1}{R^2} \iiint_V G(\xi,\theta,\phi ; \xi',\theta',\phi') \, dV' \qquad (4.27)$$

$$\alpha_2 = \frac{1}{R} \iint_{\text{sink}} G(\xi,\theta,\phi ; 1,\theta',\phi') \, dS'. \qquad (4.28)$$

The major approximation of our technique is to treat the sink as a point on the sphere and satisfy solvability and boundary condition Eq. 4.3 on the average. This assumption is reasonable for small θ_c and for points (ξ,θ,ϕ) far from the sink. An alternative would be to satisfy Eq. 4.3 only at the center of the sink by replacing β_1 and β_2 by

$$\beta'_1 = \frac{1}{R^2} \iiint_V G(1,0,0 ; \xi',\theta',\phi') \, dV' \qquad (4.29)$$

$$\beta'_2 = \frac{1}{R} \iint_{\text{sink}} G(1,0,0 \; ; \; 1,\theta',\phi') \; dS'. \tag{4.30}$$

Integrals were evaluated numerically using Gauss-Legendre quadrature. For θ_c less than 0.5 radians, the value of β_1' is within 1.2% of β_1, and the value of β_2' is within 8% of β_2. β_2' was used in calculations instead of β_2 because of its convergence properties.

One might assume that all ligand molecules dissociate from their receptors before a tubule forms or is attached to the vesicle, that is, prior to the sorting time. Therefore an average over all possible initial positions within the volume of the sphere gives the averaged mean capture time for the ligand, $\overline{W_L}$

$$\overline{\overline{W}}_L = \frac{R^2}{D_L} F_L(\theta_c), \tag{4.31}$$

where $F_L(\theta_c)$ denotes the average over the sphere of $F_L(\theta_c; \xi,\theta,\phi)$. If instead it is assumed that ligand molecules do not dissociate before a tubule forms, that is, that ligands dissociate from their receptors at the start of the sorting time, then an average of W_L over all possible initial positions on the surface of the sphere will give the average time for a ligand molecule to travel from a receptor to the tubule entrance. The notation $W_L^{(\text{surf})}$ will be used to denote this surface average.

Receptor Motion

The mean time for receptor molecules to move from the vesicle into the endosome tubule, both in the absence and presence of a membrane current flowing from the vesicle into the tubule, can be calculated. We must first derive the appropriate equation for the receptor capture time, analogous to the mean capture time equation used for the ligand (Eq. 4.1). Following the method used by Berg and Purcell (1977) for the case of molecule movement by pure diffusion, the relevant equation will be derived in a two dimensional Cartesian coordinate system. The geometry for the derivation is shown in Figure 4.3. The capture time associated with a particular position (x,y) is defined as $W(x,y)$, and the current $v(x)$ acts to move molecules in the direction of increasing x.

The movement of a molecule in the (x,y) space is modeled by a random walk on a square lattice. The capture time associated with any position (x,y) can be expressed as a time increment Δt plus the sum of the capture times associated with each of the possible new positions at which the molecule may be located after time Δt, with each of the capture times

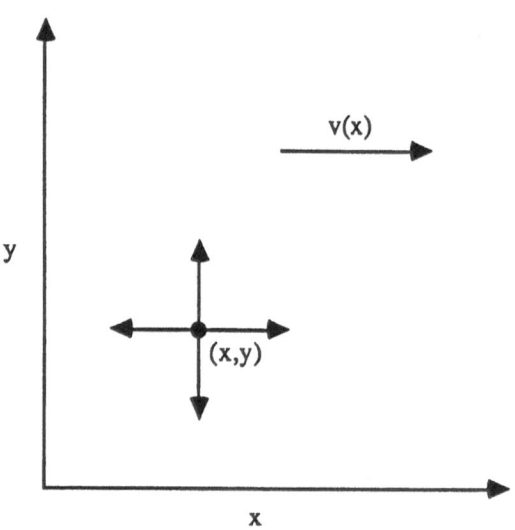

FIGURE 4.3 Geometry for derivation of mean capture time equation in the presence of a current.

weighted according to the probability that the molecule moves to that position. Pure diffusion results in a displacement of length δ in time Δt, and the effect of the current on displacement in the x direction must also be noted. Thus the difference equation for the capture time W is

$$W(x,y) = \Delta t + \frac{1}{4} [W(x+v\Delta t, y+\delta) + W(x+v\Delta t, y-\delta) +$$

$$W(x+\delta+v\Delta t, y) + W(x-\delta+v\Delta t, y)]. \tag{4.32}$$

We then make the substitution $\Delta t = \delta^2/4D$ (Einstein 1926), where D is the translational diffusion coefficient of the molecule, and take the limit of the equation as the step length δ approaches zero. The resulting equation is

$$D(\frac{\partial^2 W}{\partial x^2} + \frac{\partial^2 W}{\partial y^2}) + v(x) \frac{\partial W}{\partial x} + 1 = 0, \tag{4.33}$$

the mean capture time equation in two dimensions with a convective current $v(x)$. Note that when the current $v(x)$ is equal to zero, the mean capture time of Berg and Purcell (1977), Eq. 4.1, is recovered.

We will assume that the radius of the vesicle does not change during the sorting time, as stated earlier. To apply Eq. 4.33 to our system, then, the equation in spherical coordinates with $r = R$ is needed. Assuming that the membrane current acts only in the θ direction, the equation is

$$D_R [\frac{1}{R^2 \sin\theta} \frac{\partial}{\partial\theta} (\sin\theta \frac{\partial W_R}{\partial\theta}) + \frac{1}{R^2 \sin^2\theta} \frac{\partial^2 W_R}{\partial\phi^2}]$$

$$+ v(\theta) \frac{1}{R} \frac{\partial W_R}{\partial\theta} + 1 = 0, \tag{4.34}$$

where W_R is the capture time for the receptor and D_R is its diffusion coefficient within the membrane of the endosome. The current $v(\theta)$ is defined as positive in the direction of increasing θ coordinate. The appropriate boundary conditions are

$$W_R = 0 \qquad\qquad \theta = \theta_c \tag{4.35}$$

$$\frac{\partial W_R}{\partial \theta} = 0 \qquad \theta = \pi \tag{4.36}$$

$$W_R (\phi = 0) = W_R (\phi = 2\pi) \tag{4.37}$$

$$\frac{\partial W_R}{\partial \phi} (\phi = 0) = \frac{\partial W_R}{\partial \phi} (\phi = 2\pi). \tag{4.38}$$

There is symmetry in the ϕ coordinate; therefore, the ordinary differential equation in the independent variable θ alone can be solved to get the capture time for the receptor.

When there is no current, $v(\theta)$ is equal to zero. Eq. 4.34 can be solved analytically for this case to give

$$W_R (\theta) = \frac{R^2}{D_R} \ln \left[\frac{1 - \cos\theta}{1 - \cos\theta_c} \right] \qquad \theta \geq \theta_c \text{ and } v(\theta) = 0 \tag{4.39}$$

where θ is the initial angular position of the receptor. After averaging over all possible initial positions on the surface of the sphere, the average mean capture time $\overline{W_R}$ is obtained

$$\overline{\overline{W}}_R = \frac{R^2}{D_R} F_R (\theta_c), \tag{4.40}$$

where

$$F_R (\theta_c) = \frac{2 \ln \left[\dfrac{2}{1 - \cos\theta_c} \right]}{1 + \cos\theta_c} - 1 \qquad v(\theta) = 0. \tag{4.41}$$

Eq. 4.34 can also be solved for a nonzero value of the current $v(\theta)$; this requires that the functional form of the current be determined. We will assume for simplicity that a constant membrane area is drawn from the vesicle into the tubule per unit time and define α as the fraction of the total vesicle area that is removed from the vesicle per unit time. α is also equal to the fraction of total vesicle area that is donated to the vesicle per unit time, because we assume that the radius of the vesicle does not change during the sorting time and it has been shown that biological membranes maintain an almost constant density of molecules per area of membrane (Evans and Skalak 1980). A shell balance around a differential element of the sphere gives

$$- 2\pi R \sin(\theta + \Delta\theta)\, v(\theta + \Delta\theta) + \alpha 2\pi R^2 \sin(\theta)\, \Delta\theta = -2\pi R \sin(\theta)\, v(\theta), \tag{4.42}$$

where the first term represents membrane donated by flow into a differential element, the second term represents membrane donated by the fusion of small incoming endocytic vesicles with the sphere, and the last term represents membrane removed by flow out of the differential element. Note that $v(\theta)$ is not equal to $v(\pi - \theta)$ because the new membrane is added uniformly over the sphere; the mean capture time equation requires that all initial positions of the receptor molecule are equally likely. In the limit as $\Delta\theta$ approaches zero, the first order ordinary differential equation for $v(\theta)$ is produced,

$$\frac{d}{d\theta}\left(\sin\theta\, v(\theta) \right) = \alpha R \sin\theta, \tag{4.43}$$

with the boundary condition

$$v(\theta) = 0 \qquad \theta = \pi. \tag{4.44}$$

The solution is

$$v(\theta) = \frac{-\alpha R\,(\cos\theta + 1)}{\sin\theta} \qquad \theta_c < \theta < \pi. \tag{4.45}$$

This result was also found by Weigel (1979) for a related problem of flow on the cell surface.

For the case of a nonzero membrane current then, Eq. 4.34 must be solved with $v(\theta)$ given by Eq. 4.45 and the boundary conditions given by Eqs. 4.35 - 4.38. Analytic techniques allow only partial solution of this equation for the capture time $W_R(\theta)$,

$$W_R(\theta) = \frac{R^2}{D_R} \frac{-1}{\beta-1} \left[\int_{\theta_c}^{\theta} (\sin \frac{\theta}{2})^{2\beta-1} (\cos \frac{\theta}{2})^{-1} d\theta \right.$$

$$\left. + 2 \ln \left[\frac{\cos \frac{\theta}{2}}{\cos \frac{\theta_c}{2}} \right] \right] \qquad \theta \geq \theta_c, \, v(\theta) \neq 0, \, \beta \neq 1, \qquad (4.46a)$$

$$W_R(\theta) = \frac{-2R^2}{D_R} \int_{\theta_c}^{\theta} \frac{\sin \frac{\theta}{2}}{\cos \frac{\theta}{2}} \ln(\sin \frac{\theta}{2}) d\theta \qquad \theta \geq \theta_c, \, v(\theta) \neq 0, \, \beta = 1, \qquad (4.46b)$$

where the dimensionless current β is defined by

$$\beta = \frac{\alpha R^2}{D_R} . \qquad (4.47)$$

To calculate the average value of the receptor capture time, $W_R(\theta)$ is averaged over all possible initial positions of the receptor on the surface of the sphere, $\theta_c < \theta < \pi$, to obtain

$$\ddot{W}_R = \frac{R^2}{D_R} F_R(\theta_c, \beta), \qquad (4.48)$$

where

$$F_R(\theta_c, \beta) = (\frac{1}{1+\cos \theta_c})(\frac{-1}{\beta-1}) \left[\int_{\theta_c}^{\pi} \int_{\theta_c}^{\theta'} (\sin \frac{\theta}{2})^{2\beta-1} (\cos \frac{\theta}{2})^{-1} d\theta \sin \theta' d\theta' \right.$$

$$\left. - (1+\cos \theta_c) \right] \qquad v(\theta) \neq 0, \, \beta \neq 1 \qquad (4.49a)$$

$$F_R(\theta_c,\beta) = \left(\frac{-2}{1+\cos\theta_c}\right) \int\int_{\theta\theta_c}^{\pi\theta'} \frac{\sin\frac{\theta}{2}}{\cos\frac{\theta}{2}} \ln\left(\sin\frac{\theta}{2}\right) d\theta \sin\theta' \, d\theta'$$

$$v(\theta) \neq 0, \; \beta = 1. \tag{4.49b}$$

The values of the integrals in Eqs. 4.46 and 4.49 were computed numerically by Gauss-Legendre quadrature.

4.3 Results

The average time required by ligand molecules to reach the tubule entrance is given in Eq. 4.31; the average times for receptor molecules in the absence or presence of a convective membrane current are given Eqs. 4.40 and 4.48, respectively. Each is expressed as the product of the ratio R^2/D_n and a factor F, the dimensionless time, which depends only on the angle θ_c and, in the case of receptor molecule diffusion in the presence of a current, the dimensionless current β.

The variations of the dimensionless times F_R ($\beta = 0$) and F_L with θ_c are shown in Figure 4.4. As previously discussed by Adam and Delbrück (1968), this "tracking factor" F reflects the dimensionality of the system: the two dimensional F_R varies as $\log(1/\theta_c)$ and the three-dimensional F_L as $1/\theta_c$ for small θ_c. Because of this difference in dimensionality, the dimensionless capture time for the ligand is always greater than that for the receptor. As the angle θ_c decreases, both dimensionless times increase and the difference between the two also increases.

Examination of the numerical results for $F_L(\theta_c)$ indicates that as θ_c decreases, the term $2\beta_2'/(3(1-\cos\theta_c))$ dominates Eq. 4.26 (substituting β_2' for β_2) and that β_2' can be approximated by θ_c. Together with the small angle approximation for $\cos\theta_c$, this implies that

$$\overset{...}{W}_L \cong \frac{R^2}{D_L} \frac{4}{3\theta_c}. \tag{4.50}$$

This approximation is good to within 4% for $\theta_c < 0.1$ radians and to within 1% for $\theta_c < 0.01$ radians.

We also calculated the surface-averaged capture time for the ligand, W_L(surf). The results of this calculation are differ by only a few percent from the volume-averaged $\overline{W}_L$ discussed above and are not shown here for that reason.

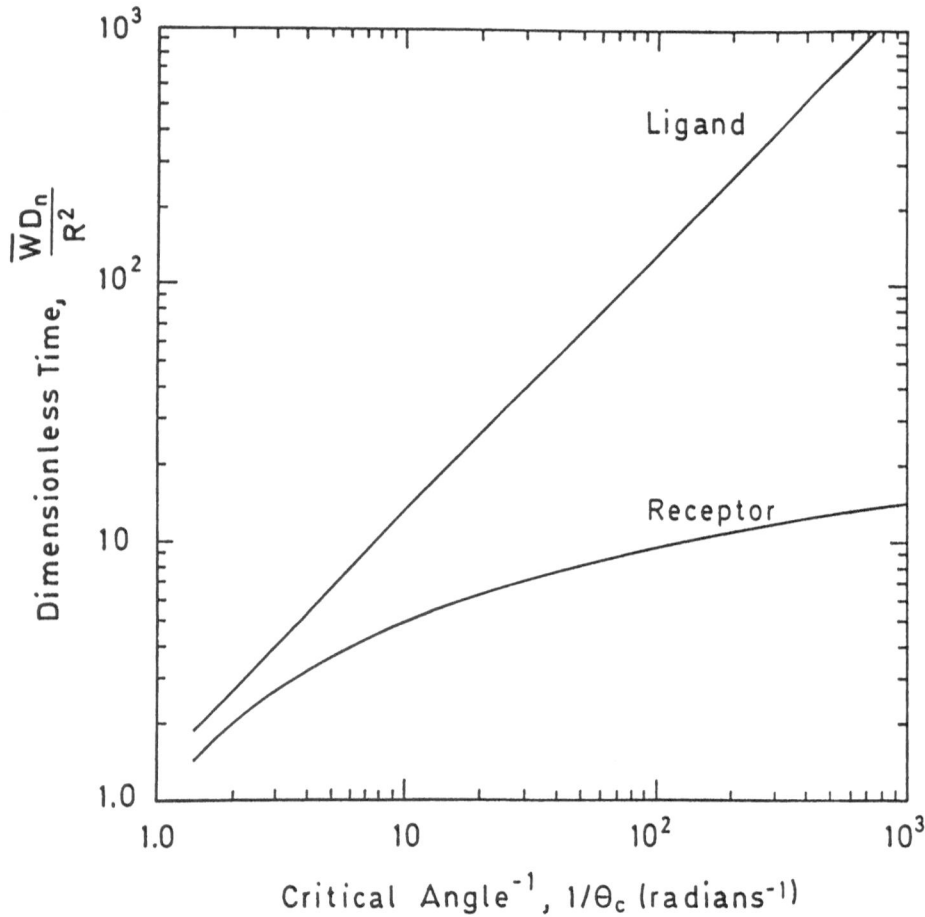

FIGURE 4.4 Dimensionless times for ligand and receptor molecules to reach the tubule entrance as a function of the critical angle θ_c. The dimensionless current β is equal to 0.

The variation of the dimensionless time F_R ($\beta \neq 0$) with θ_c is shown in Figure 4.5. Clearly, the addition of a current to carry receptors into the tubule reduces the receptor capture time. For small values of the dimensionless current β, diffusion is the primary mechanism for moving receptors into the tubule. As the current increases in magnitude, convection plays an increasing role in this movement of receptors until at high values of β convection is the primary mechanism.

To compare the average times required by the molecules to reach the sink, we calculate the ratio of the capture times

$$\frac{\overline{\overline{W}}_R}{\overline{\overline{W}}_L} = \frac{D_L}{D_R} \frac{F_R(\theta_c, \beta)}{F_L(\theta_c)} , \tag{4.51}$$

which is independent of the vesicle radius R and a function only of the angle θ_c, the dimensionless current β, and the ratio of diffusion coefficients. In Figure 4.6, lines of constant $\overline{W}_R(\theta_c, \beta) / \overline{W}_L(\theta_c)$ are shown as a function of θ_c^{-1} and D_R/D_L for $\beta = 0$. It is clear that for a constant ratio of diffusion coefficients, the receptor's advantage of searching for the tubule entrance in one less dimension than the ligand is greatest at small θ_c. In other words, for small enough θ_c, the mean time for the receptor to reach the tubule entrance can be less than that for the ligand ($\overline{W}_R(\theta_c, \beta) / \overline{W}_L(\theta_c) < 1$) even though the ligand's diffusion coefficient is orders of magnitude greater than that for the receptor.

When the value of the dimensionless current β is greater than zero, the receptor gains a further advantage over the ligand. This is shown in Figure 4.7. As the value of the current increases, the angle at which the receptor's capture time is less than the ligand's capture time is increased over the case of $\beta = 0$ for the same ratio of diffusion coefficients.

4.4 Discussion

Mathematical Analysis

In this chapter, we have solved mean capture time equations in two and three dimensions with spherical geometry. The two-dimensional problem is straightforward, but the three-dimensional problem requires some care because it involves a mixed boundary condition. Our approach to this latter problem is guided by the work of other investigators on a related problem, the calculation of the rate constant for diffusion of ligand to cell surface receptors (Berg and Purcell 1977; DeLisi and Wiegel 1981; Brunn 1981). This problem requires the

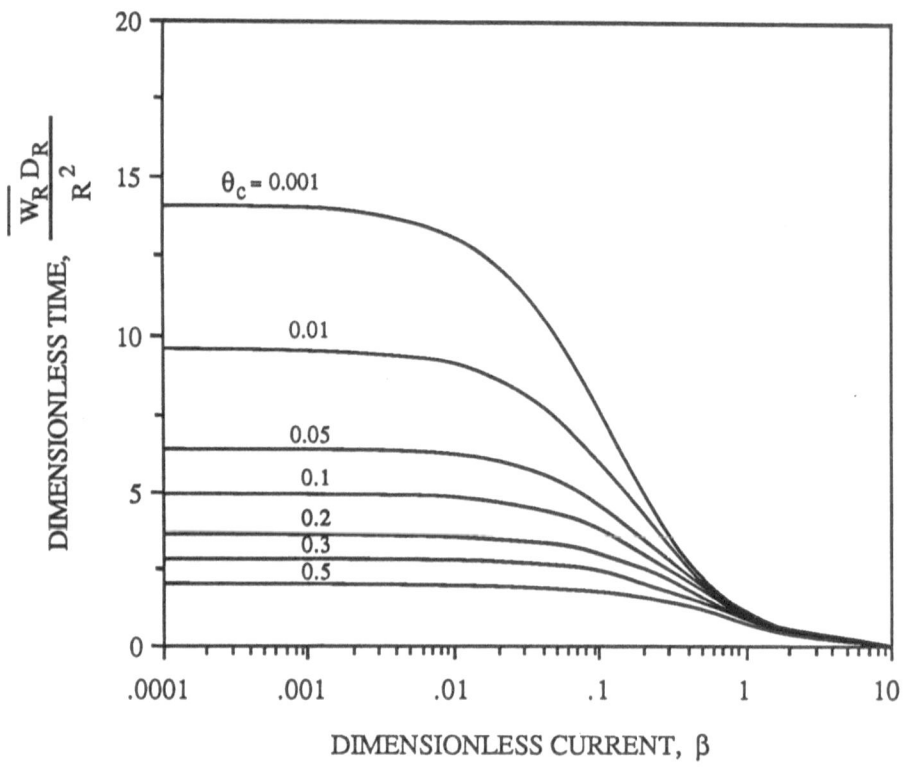

FIGURE 4.5 Dimensionless time for receptor molecules to reach the tubule entrance as a function of the critical angle θ_c and the dimensionless current β.

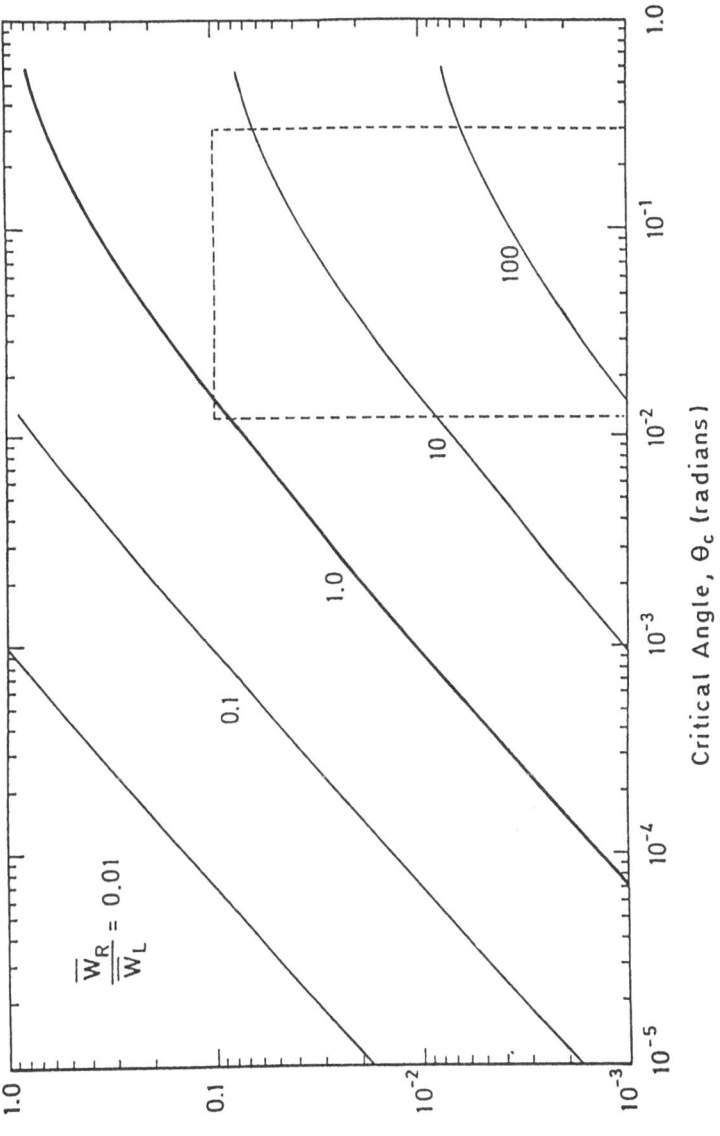

FIGURE 4.6 Ratio of the receptor capture time with no current, $\overline{W}_R(\beta = 0)$, to the ligand capture time, $\overline{W}_L$, as a function of the ratio of diffusion coefficients and the critical angle. The curve $\overline{W}_R/\overline{W}_L = 1$ separates two regimes: when $\overline{W}_R/\overline{W}_L < 1$, receptors are able to find the tubule entrance more rapidly than ligand molecules and when $\overline{W}_R/\overline{W}_L > 1$, ligand molecules are able to find the tubule entrance more rapidly than receptors. Estimated parameter values lie within the boxed region; the box extends to a diffusion coefficient ratio of 10^{-6} (not shown).

44

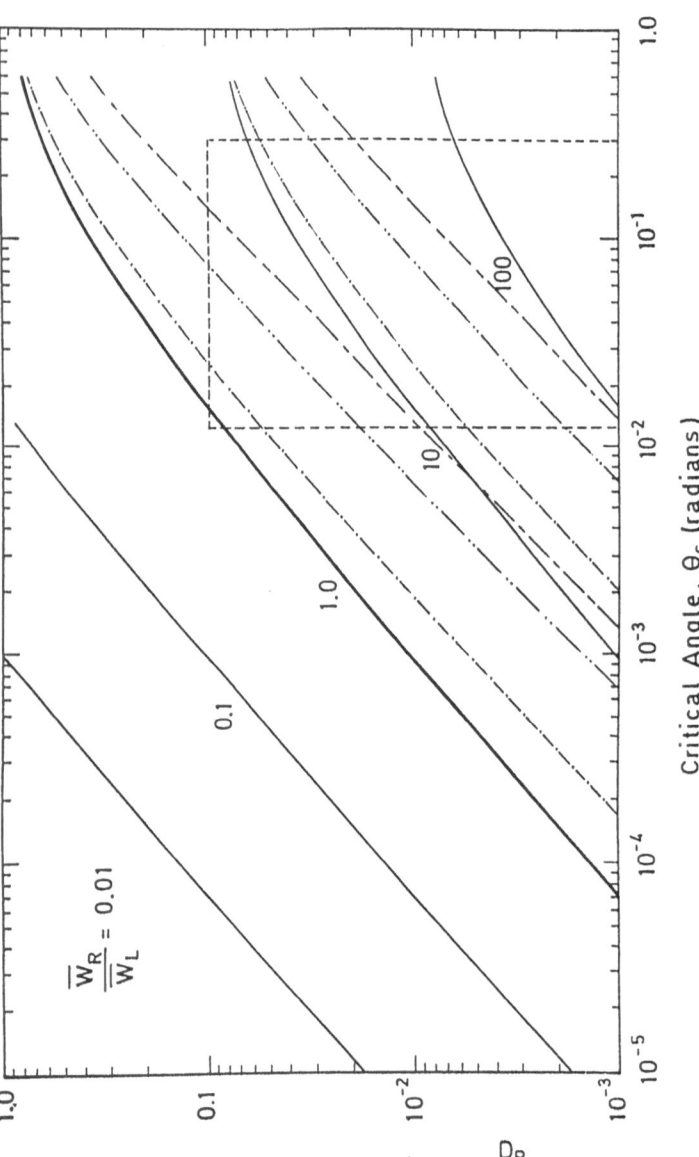

Critical Angle, Θ_c (radians)

Ratio of Diffusion Coefficients, $\dfrac{D_R}{D_L}$

FIGURE 4.7 Ratio of the receptor capture time with current, $\overline{W}_R$, to the ligand capture time, $\overline{W}_L$, as a function of the ratio of diffusion coefficients, the critical angle, and the dimensionless current. Line markings: ——— $\beta = 0$; — ·· — $\beta = 0.1$; — · — $\beta = 0.5$; — — — $\beta = 1.0$. Shown are the shift in the $\overline{W}_R/\overline{W}_L = 1$ (upper three curves for which $\beta \neq 0$) and $\overline{W}_R/\overline{W}_L = 10$ (lower three curves for which $\beta \neq 0$) curves. As in Fig. 4.6, the curve $\overline{W}_R/\overline{W}_L = 1$ separates the regime in which receptors are able to find the tubule entrance more rapidly than ligand molecules ($\overline{W}_R/\overline{W}_L < 1$) and the regime in which ligand molecules are able to find the tubule entrance more rapidly than receptors ($\overline{W}_R/\overline{W}_L > 1$). Estimated parameter values lie within the boxed region; the box extends to a diffusion coefficient ratio of 10^{-6} (not shown).

solution of the diffusion equation with mixed boundary conditions. These boundary conditions result from the situation where part of the surface is covered with absorbers (receptors) while the remainder of the surface is reflecting. Fortunately, one can postulate a constant ligand concentration infinitely far from the cell and thus solve the steady state diffusion equation. To solve the equation describing this problem, Laplace's equation on the exterior of a sphere, several methods have been used. In particular, the method of Brunn (1981) treats the absorbing boundary condition as an unknown inhomogeneity in flux and solves the problem approximately using Green's functions. Shoup et al. (1981) also solve for an unknown flux at a sink and use this in obtaining rate constants for different geometries.

To find the rate constant for ligand entry into a tubule, the endosome problem, the solution to the diffusion equation within a sphere with mixed boundary conditions on the surface of that sphere is needed. Because there is no infinite source of ligand for this interior problem, the equation cannot be solved at steady state. This complication is avoided by instead using the mean capture time equation, derived by Berg and Purcell (1977) and Szabo et al. (1980), to calculate the time required for a ligand molecule to reach the tubule entrance. If the tubule entrance can be considered perfectly absorbing, then this mean time — averaged over the volume or surface area of the sphere to account for all possible initial positions of ligand or receptor molecules — is simply the inverse of the rate constant sought. If, however, the entrance is only partially absorbing, this first rate constant for finding the entrance can be combined with a second rate constant for adsorption or reaction, analogous to the "encounter complex" model of receptor-ligand binding (DeLisi and Wiegel 1981; DeLisi 1980).

Thus we solve the mean time equation, Poisson's equation inside a sphere with mixed boundary conditions on the surface. To do this, an approximate Green's function technique is used. Following Brunn (1981), we treat the absorbing boundary condition as an unknown inhomogeneity in flux. A volume integral is used to account for the inhomogeneity in the equation itself. The result can be expressed as the product of a ratio, R^2/D_L, and a factor $F_L(\theta_c)$, which is plotted in Figure 4.4. For small θ_c, we obtain the approximate result given in Eq. 4.50.

This result can be compared with the results of similar calculations in the literature, as shown in Figure 4.8, in order to discern the influence of geometry on the ability of a ligand to find a generalized surface reaction site. The diffusive current of ligand to a perfectly absorbing circular cell surface receptor is equivalent to the diffusion of molecules to a circular sink on an infinite plane in the limit of an infinitely small receptor. As found by Hill (1975) and by Berg and Purcell (1977), the rate constant k is equal to 4Db where D is the diffusion coefficient and b is the radius of the sink. Shoup and Szabo (1982) show that the first passage time, the quantity we refer to as $\overline{W_L}$, is equal to a volume divided by this rate constant. The appropriate volume is that of our vesicle, $4\pi R^3/3$, and for the same size sink we set b equal to $R\theta_c$. Thus

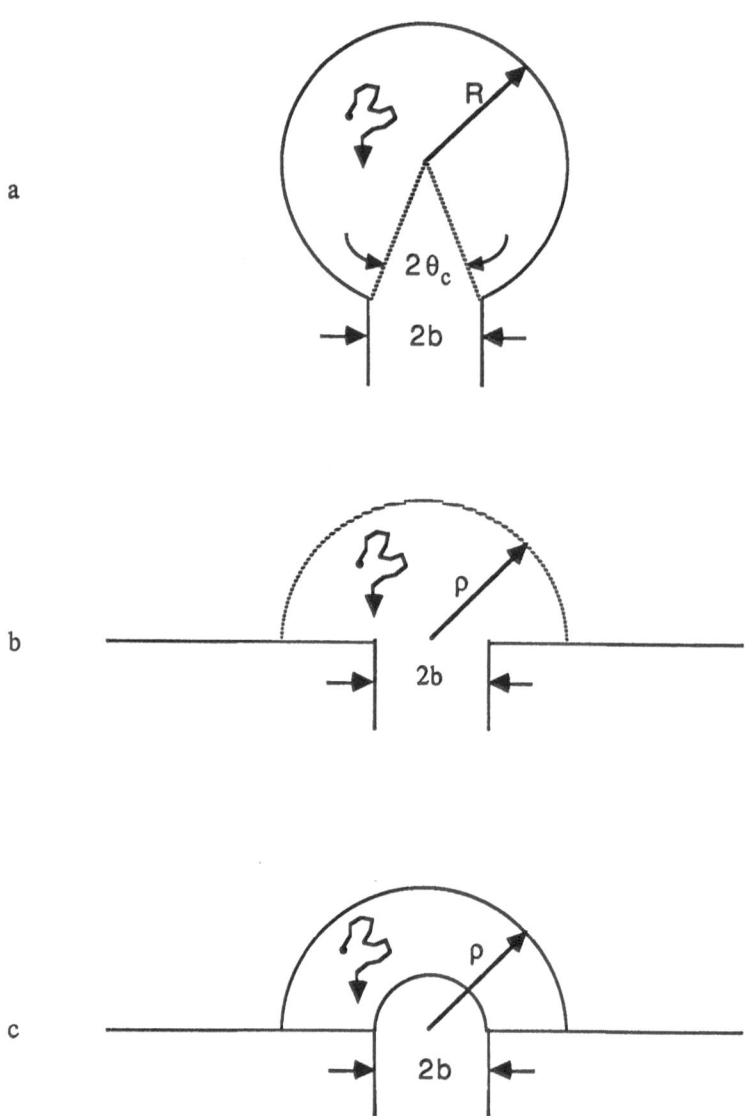

FIGURE 4.8 Geometries for comparison of ligand capture times. (a) Diffusion within a sphere to a hole on the surface. (b) Diffusion to a hole on an infinite plane. (c) Diffusion within a hemisphere to a smaller concentric hemisphere.

the mean time $\overline{W}_L{}^{FP}$ for diffusion to a sink on a flat plate is equal to

$$\overline{W}_L^{FP} = \frac{R^2}{D_L}\frac{\pi}{3\theta_c}. \tag{4.52}$$

$\overline{W}_L{}^{FP}$ is smaller than our result because of geometrical differences between the problems. In our problem, shown in Figure 4.8a, ligand moves within a sphere to a hole on the surface of that sphere. As shown in Figure 4.8b, the volume used for diffusion to a flat plate should be placed so that the hole is most accessible to the ligand: that is, as a hemisphere that has as its center the hole. Although the radius ρ of this hemisphere is greater than R in order that the hemisphere volume is equal to the volume of a sphere with radius R, the average distance traveled to the sink is less than in Figure 4.8a. So it is clear that our result should be greater than the result for diffusion to a hole on a flat plate for purely geometric reasons.

Second, we compare our result with that of Adam and Delbrück (1968), as simplified considerably by Szabo et al. (1980), who calculated the mean time for a molecule within a hemisphere of radius ρ to diffuse to the surface of a smaller concentric hemisphere of radius b (see Figure 4.8c). When $\rho >> b$, their results are well approximated by a time of $\rho^2/3D\theta$ where θ equals b/ρ. To equate the volume of this outer hemisphere with our sphere and the sink radius with our radius b, we set ρ equal to $2^{1/3}R$ and b equal to $R\theta_c$. Thus Adam and Delbrück's result becomes

$$\overline{W}_L^H \cong \frac{R^2}{D_L}\frac{2}{3\theta_c}, \tag{4.53}$$

which is smaller than Berg and Purcell's result because the inner hemisphere is slightly easier to find than the hole of Figure 4.8b. Our result is the largest of the three, illustrating the effect of geometry. In Figure 4.9, we plot the three results over a range of sink sizes.

The effect of geometry is also demonstrated by the influence of dimensionality on the transport rate constants. The receptor searches for the tubule entrance in one fewer dimension than the ligand, and this dimensionality advantage can outweigh the ligand's advantage of a greater diffusion coefficient when the critical angle is small enough.

Biological Implications

Because an equilibrium mechanism does not explain the observed intracellular

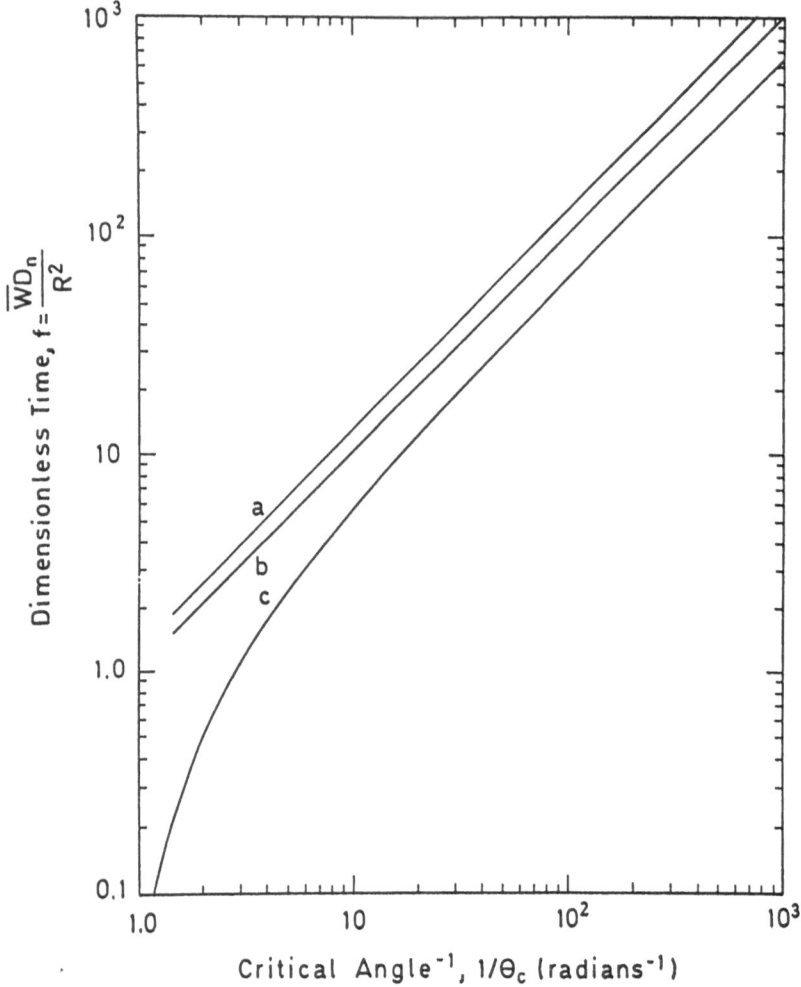

FIGURE 4.9 Effect of geometry on ligand capture times. Curves correspond to the geometries shown in Fig. 4.8. In all cases, $\theta_c = b/R$. (a) Diffusion within a sphere to a hole on the surface. Our result is replotted from Fig. 4.4. (b) Diffusion to a hole on an infinite plane. For $\rho = 2^{1/3}R$, $\overline{W}_L{}^{FP}D_L/R^2 = \pi/3\theta_c$. (c) Diffusion within a hemisphere to a smaller concentric hemisphere. For $\rho = 2^{1/3}R$ and $x = 2^{-1/3}\theta_c$, Szabo, et al. (1980) give the dimensionless capture time as

$$\frac{\overline{W}_L{}^H D_L}{R^2} = \frac{2^{2/3}(1-x)^2(5+6x+3x^2+x^3)}{15x(1+x+x^2)}.$$

segregation of receptor and ligand molecules within the endosome, the sorting chamber for the cell, two kinetic schemes for the separation are suggested. First, we propose that receptors may be able to move into endosome tubules from the vesicular region of the endosome more rapidly than ligand molecules due to an advantage in dimensionality. This dimensionality advantage may be in addition to the enhanced rate of transport into the tubule that receptors experience if a convective membrane current is present. Alternatively, we consider the possibility that receptors diffusing into tubules are trapped there whereas ligand molecules are free to diffuse back out of the tubule. The evaluation of these possible sorting mechanisms requires the calculation of the rates at which receptor and ligand molecules move into endosome tubules.

Beginning with a simple model of the endosome, a single vesicle with an attached thin tubule, we calculate the capture time, or time to reach the tubule entrance, for molecules placed uniformly throughout the vesicle and moving by passive diffusion alone or by diffusion plus convection. The calculated capture times for the receptor and ligand are a function of four parameters of the system: the vesicle radius, the tubule radius, the appropriate diffusion coefficient, and, in the case of the receptor, the dimensionless current. To compare the receptor and ligand capture times, the ratio of times $\overline{W_R}(\beta=0) / \overline{W_L}$ as a function of the ratio of diffusion coefficients D_R/D_L and the ratio of the tubule radius to the vesicle radius b/R (or θ_c) is plotted in Figure 4.6. In Figure 4.7, $\overline{W_R}(\beta\neq0) / \overline{W_L}$ is shown. The curve of parameter values for which the ratio $\overline{W_R}/\overline{W_L}$ is equal to one separates two regimes in the figures. When the value of the ratio is greater than one, the transport of receptor to the tubules is slower than the transport of ligand molecules, and when the value is less than one, receptors diffuse to the tubule entrance more rapidly than ligand. Using the appropriate parameter values, the regime in which the endosome is likely to operate can be determined.

From electron micrographs, Geuze et al. (1983) and Marsh et al. (1986) obtain values for the vesicle radius R and tubule radius b. The parameter ranges and the calculated range of values for θ_c are given in Table 4.1. The diffusion coefficients D_R and D_L within an endosome are not known, so we postulate that they are not significantly different from the diffusion coefficients for receptors on the cell surface and for ligand free in solution. Thus we estimate that D_L is 10^{-7} to 10^{-5} cm^2/sec and D_R is 10^{-11} to 10^{-8} cm^2/sec (Wiegel 1984; Menon et al. 1986a; Schlessinger et al. 1978; Hillman and Schlessinger 1982; Giugni et al. 1987). Assuming these parameter values to be accurate, one can determine from the region inside the dashed lines in Figure 4.6 that $\overline{W_R}(\theta_c)$ is likely 10 to 100 times greater than $\overline{W_L}(\theta_c)$ when there is no membrane current to aid receptors in their entry into the tubule. Thus the endosome would operate in the regime for which $\overline{W_R}/\overline{W_L}$ is greater than one, implying that the cell cannot sort the receptors into a tubule before a significant amount of ligand reaches the tubule if β is equal to zero. In fact, these parameter values indicate that ligand molecules will

Model Parameter	Range
Vesicle radius, R	10^{-5} - 4×10^{-5} cm.
Tubule radius, b	5×10^{-7} - 3×10^{-6} cm.
Critical angle, θ_c	0.0125 - 0.30 radians
Receptor diffusion coefficient, D_R	10^{-11} - 10^{-8} cm^2/sec
Ligand diffusion coefficient, D_L	10^{-7} - 10^{-5} cm^2/sec

TABLE 4.1 Parameter values. Values for R and b are taken from Geuze et al. (1983) and Marsh et al. (1986) and are used to calculate θ_c. Diffusion coefficients are estimated as described in the text.

have equilibrated throughout the vesicle and tubule volumes by the time an appreciable number of receptors have diffused into the tubule. Thus the analysis reveals that our first proposed separation scheme is unlikely when there is no membrane current; the ligand's advantage of a greater diffusion coefficient prevails over the receptor's advantage in dimensionality.

If value of the dimensionless current β is greater than zero, the receptor capture time is decreased and the ratio of times $\overline{W_R}/\overline{W_L}$ is decreased. Thus for the same reasonable ranges of the parameter values D_R, D_L, and θ_c, the ratio of times $\overline{W_R}/\overline{W_L}$ is more likely to be less than one, implying receptors may be able to move into the tubule more rapidly than ligands. This is shown in Figure 4.7. It is important, therefore, to estimate a reasonable value for the current β. Since it is not known whether a membrane current exists in the endosome, there is no measure of a membrane velocity there. On the cell surface, however, membrane velocities have been measured. It is possible that the magnitude of the membrane current inside the endosome, were it to exist, is similar to the magnitude of currents on the cell surface. Bretscher (1976, 1984) and Abercrombie et al. (1970) found that a reasonable estimate for this current is on the order of 1 to 3 μm/min. If we assume that this velocity is approximately equal to the average velocity v on the sphere surface $(0 \leq \theta \leq \pi)$, $\alpha\pi R/2$, and we use the ranges of R and D_R listed in Table 4.1, we can calculate the range of β values that might be present in the endosome from

$$\beta = \frac{2\bar{v}R}{\pi D_R}.$$

$$(4.54)$$

For example, for v = 1 to 3 μm/min., R = 1 x 10^{-5} to 4 x 10^{-5} cm., and $D_R = 10^{-11}$ to 10^{-8} cm^2/sec, β ranges from 0.0011 to 13. The diffusion coefficient of cell surface receptors is most likely on the order of 10^{-10} cm^2/sec (Menon et al. 1986a; Schlessinger et al. 1978; Hillman and Schlessinger 1982; Giugni et al. 1987), and if we use this value instead of the range for D_R given above we obtain a narrower range for β of 0.11 to 1.3.

From Figure 4.7 it is apparent that while a current of $\beta = 1$ would increase the probability that the endosome operates in the regime for which $\overline{W_R}/\overline{W_L}$ is less than one and sorting based on a difference in capture times is possible, the chance of this occurring is still small. For this separation scheme to operate well, the ratio $\overline{W_R}/\overline{W_L}$ would have to be significantly less than one, this is not true over nearly all of the parameter space that we feel is reasonable. Note particularly that for the most reasonable value of the receptor diffusion coefficient, D_R on the order of 10^{-10} cm^2/sec, the ratio D_R/D_L is less than or equal to 10^{-3}. In this region of the parameter space, the dimensionless current β would need to be much greater than we have estimated to make possible a separation of receptors and ligands. Thus we feel that sorting based on a difference in capture times is an unlikely mechanism for the separation of receptor and ligand molecules, even in the presence of a small membrane current to enhance receptor motion.

It is possible, however, that the parameter values we have used are inaccurate and that receptors do reach the tubule more rapidly than ligand molecules. The tubule radius may be much smaller at the entrance than where it is measured, further down the tubule. Such a constriction would decrease θ_c considerably and greatly increase the receptor's dimensionality advantage over the ligand. For example, if the tubule radius at the entrance is constricted to 25 Å, the vesicle radius is 4×10^{-5} cm, and D_R/D_L is equal to 0.1, then the receptor would move into the tubule about twice as fast as the ligand in the absence of any membrane current. At such small tubule diameters, the finite size of the ligand becomes important. The ligand effectively sees a smaller θ_c than the ratio b/R and this will further decrease the ratio $\overline{W_R}/\overline{W_L}$. In addition, the diffusion coefficients may have been inaccurately estimated. To our knowledge, no one has measured diffusion coefficients with an endosome, and D_R may be enhanced over its plasma membrane value or D_L may be less than its value in free solution. Finally, our estimated value for β may be much smaller than the true current in the endosome. All of these possible changes in the values of the parameters would decrease the ratio $\overline{W_R}/\overline{W_L}$.

At present, however, we believe that the model assumptions and parameter values are reasonable. The cell apparently cannot sort receptor and ligand molecules by taking advantage of a difference in the dimensionality of their diffusion processes.

As a second possible sorting mechanism, we suggested that receptors diffusing into tubules are trapped there. The trapping mechanism may be the interaction of receptors, particularly the part of the receptor molecule which extends through the membrane and into the cytoplasm, with trapping molecules such as clathrin or elements of the cytoskeleton. Alternatively, if a membrane current exists, it could act to trap receptors in the tubule. Weigel (1980) examined a similar problem, that of flow on the cell surface toward a sink at one end of the cell, and found that all receptors will be trapped at the sink at equilibrium if $\beta \geq 1$. For smaller values of the current, one would expect a similar though less dramatic trapping effect. This trapping effect due to the presence of a membrane current may be even greater than one would expect based only on the magnitude of β because as the tubule grows or lengthens, receptors attempting to exit the tubule will not only have to diffuse against the current but will have to travel the ever-increasing length of the tubule.

For this second sorting mechanism, the length of time a vesicle and tubule must remain connected in the endosome in order to allow the movement of a substantial fraction of receptors into the tubule can be estimated from the calculated rate constant for the movement of free receptors into a tubule. Receptors not diffusing into a tubule within this sorting time might then be degraded in lysosomes with the ligand molecules. Assuming a first order process

(Szabo et al. 1980), the fraction of receptors in the tubule as a function of time, $1-f_R(t)$, can be found simply from

$$1 - f_R(t) = 1 - \exp\left(\frac{-t}{\overline{W}_R}\right). \tag{4.55}$$

This is the fraction of receptors recycled; f_R is the fraction of receptors degraded. For a reasonable choice of parameter values, a vesicle radius of 3×10^{-5} cm, a tubule radius of 100 Å, and a diffusion coefficient of 10^{-10} cm^2/sec, one can calculate that the mean time $\overline{W}_R$ is equal to 1.1 min. and that 99% of the receptors will have moved into the tubule by 6 min. Thus, we have a means of estimating the time required to sort the two populations after the acidic environment of the endosome has released the ligand from the receptor.

Although an exact measurement of the sorting time for a particular system is not available, several related measurements do give estimates of that time. As detailed in Chapter 2, a reasonable estimate of the sorting time in many cells is 5 to 10 min. This estimate of the sorting time agrees well with our estimate in the previous paragraph of the length of time required for an appreciable number of receptors to move into the tubule, and thus supports the plausibility of this second sorting mechanism.

Because ligand molecules apparently equilibrate throughout the vesicle and tubule volumes within this sorting time, we suggest that a fraction of the ligand is "mis-sorted" into the tubule with the recycling receptors and will be exocytosed. The fraction of ligand in the tubule at equilibrium can be calculated from the volumes of each compartment and knowledge of the partition coefficient κ (Eq. 3.1). For example, asialoorosomucoid (ASOR), a ligand for the asialoglycoprotein receptor, has an approximate diameter of 40 Å[1] and, for a tubule radius of 100 Å, the partition coefficient is equal to 0.64. The fraction of ligand recycled, $1-f_L$, can then be found from

$$1 - f_L = \frac{\kappa V_T}{V_V + \kappa V_T}, \tag{4.56}$$

where V_V is the vesicle volume and V_T is the tubule volume. If the tubule volume is 30% of

---

[1] Molecular weight of ASOR is ~ 40 kD. Molecular diameter is estimated from the molecular weights and diffusivities of macromolecules of similar size and the Stokes-Einstein relation (Tanford 1961).

the total endosome volume, in the range measured by Marsh et al. (1986), then 22% of the ligand would be found in tubules at equilibrium. Assuming that the tubule becomes the recycling vehicle without losing this ligand, this is the fraction of ligand that would be exocytosed. We suggest, then, that the cell allows this small fraction of ligand to be returned to the cell surface in enabling the receptors to recycle. Although there are other cellular mechanisms, such as reversible pinocytosis (Daukas et al. 1983), that can result in exocytosis of ligand, we believe that this "mis-sorting" also may provide a mechanism for returning ligand to the cell surface.

Such exocytosis has been detected by a number of researchers. Mellman et al. (1984) found that approximately 50% of an internalized Fab ligand to the Fc receptor on macrophages returned to the cell surface. Similarly, Marshall (1985a) found exocytosis of about 25% of internalized insulin, and Townsend et al. (1984) found exocytosis of about 50% of some internalized asialoglycoproteins. Thus the predicted exocytosis from our proposed sorting mechanism has been detected in several receptor systems.

The maximum fractional amount of exocytosis our sorting mechanism allows is equal to the fraction of endosome volume found in the tubules. The measurements of Marsh et al. (1986) estimate a fractional tubule volume of 30 to 40%. If these measurements are correct for the Fc and asialoglycoprotein receptor systems mentioned above, there is a shortcoming to our model: the exocytosis it predicts is exceeded by some of the experimental data.

4.5   Conclusions

In this chapter, two possible sorting mechanisms were evaluated. The first, in which receptors are able to move more rapidly than ligands into tubules because of a dimensionality advantage and the possible existence of a membrane current, was found to be possible for some parameter ranges but not for the parameter values that most likely describe the endosome. It was also found that in the time required for an appreciable number of receptors to move into the tubule, the ligand molecules will have equilibrated between the vesicle and tubule volumes of the endosome; this latter result was used in the evaluation of the second sorting mechanism.

The second mechanism was found to give a good separation of receptors and ligands. Receptors are assumed to be trapped in the tubule once they reach it, and ligand molecules are allowed to move between the vesicle and tubule and equilibrate between these volumes. From the calculated rate constant for receptor movement and experimental estimates of the sorting time, it was shown that this mechanism will allow for large amounts of receptor recycling within the sorting time and that a fraction of the ligand will also be exocytosed.

Thus the second mechanism can explain the efficiency and kinetics of many experimentally described separations of receptors and ligands and meets the minimal requirements of a sorting mechanism that were described in Chapter 2. However, there are shortcomings to even this mechanism. Ligand exocytosis has been reported to fall within the ranges we would predict, but also in amounts greater than that range. In addition, new data on ligand effects on sorting suggest that ligand affinity and valency are factors that need to be included in a model of the sorting process; these data will be explained more completely in Chapters 5 and 6.

For these reasons, the second sorting mechanism proposed and examined here must be expanded to include these effects and account for these data.

CHAPTER 5
A WHOLE CELL KINETIC MODEL OF THE ENDOCYTIC CYCLE:
FOCUS ON SORTING

5.1 Introduction

In Chapters 3 and 4, equilibrium and kinetic schemes were examined in order to elucidate the mechanism by which receptors and ligands are sorted in the endosome. The most successful of these mechanisms, in terms of explaining the efficient recycling of receptors from the endosome back to the cell surface, suggested that receptors diffuse from the endosomal vesicles into tubules and are trapped there and that during this time ligand molecules equilibrate between the vesicular and tubular volumes. During the sorting time of 5 to 10 minutes, it is likely that 95 to 100% of the receptors reach the tubules. Thus the diffusion with trapping mechanism meets the minimal requirements for a sorting mechanism that were derived in Chapter 2.

The diffusion with trapping mechanism predicts that the maximum amount of ligand recycled or exocytosed to the cell surface is equal to the fraction of the endosome volume found in the tubules, estimated to be about 30 to 40%. As listed in Chapter 4, several systems have recently been identified in which a greater amount of ligand exocytosis has been observed (Mellman et al. 1984; Townsend et al. 1984). The diffusion with trapping mechanism does not predict these results.

In addition, it has recently been reported that the kinetics of the endocytic cycle and possibly the outcome of the sorting step itself in a particular receptor system may vary with the identity of the ligand (Hopkins and Trowbridge 1983; Weissman et al. 1986; Anderson et al. 1982; Townsend et al. 1984; Schwartz et al. 1986). For example, Mellman and coworkers (Mellman et al. 1984; Mellman and Plutner 1984; Ukkonen et al. 1986) found that the number of cell surface Fc receptors on J774 cells (a macrophage cell line) decreased dramatically upon exposure of the cells to a multivalent ligand but that there was no decrease in surface Fc receptors when the cells were allowed instead to internalize a monovalent ligand. This decrease in cell surface receptor number was shown to be the result of an enhanced rate of receptor degradation. This observed ligand effect may be the result of alterations in the rate constants for steps of the endocytic cycle not related to sorting and/or variation in the outcome of the sorting process itself.

It is important to note that the diffusion with trapping sorting mechanism does not predict these ligand effects. One would not expect the geometry of the endosome, the sorting time, the receptor diffusion coefficient, or the membrane current to be affected by the ligand identity. Although the ligand identity would determine the value of the partition coefficient, changes in this parameter value alone are not expected to have a significant effect on the sorting outcome and therefore the cycle kinetics. A change only in the value of the partition coefficient, for example, cannot predict the increased receptor degradation observed in the macrophage Fc receptor system mentioned above.

Thus an analysis of data on the effects of ligands on the dynamics of the endocytic cycle is expected to lead to one of two conclusions. First, the ligand effects may be due to alterations in steps in the cycle other than sorting. For example, the ligand identity might affect the rate of internalization by some as yet undetermined mechanism. Second, the ligand effects on the cycle may be shown to be a result of a variation in the sorting outcome with the ligand identity. As discussed above, the diffusion with trapping mechanism is not able to predict these changes in the sorting outcome. In this case, then, a new sorting mechanism should be proposed.

In this chapter, we develop a model for the entire endocytic cycle and focus on the characterization of the sorting step. The outcome of sorting is described by the sorting fractions introduced in Chapter 4, f_R and f_L, the fractions of receptors and ligands slated for degradation (or fates other than recycling) at the end of the sorting step. This whole cell kinetic model is developed so that it can be used to extract information about the outcome of the sorting process, specifically, the values of the sorting fractions f_R and f_L, from whole cell experimental data. These values can then be compared with the sorting fractions predicted by the diffusion with trapping mechanism or any other proposed sorting mechanism so that the success of the proposed sorting mechanism in describing the experimental data can be assessed.

5.2 Whole cell kinetic model

A detailed analysis of the sorting process is best understood in the context of a whole cell kinetic analysis of receptor and ligand traffic during endocytosis. As shown in Figure 5.1, rate constants can be assigned to each of the steps in the cycle. Binding and dissociation of a ligand and its receptor at the cell surface are assigned the rate constants k_1^N and k_{-1}^N, where the superscript N refers to the value of these rate constants at the normal extracellular pH. If the ligand has more than one binding site for the receptor, additional rate constants must be assigned for the crosslinking reaction(s), as shown in Figure 5.2. The term crosslinking in this context refers to the bridging of several receptors molecules by a ligand binding to each of the receptors simultaneously. The rate constant for the internalization of the complexes is k_I;

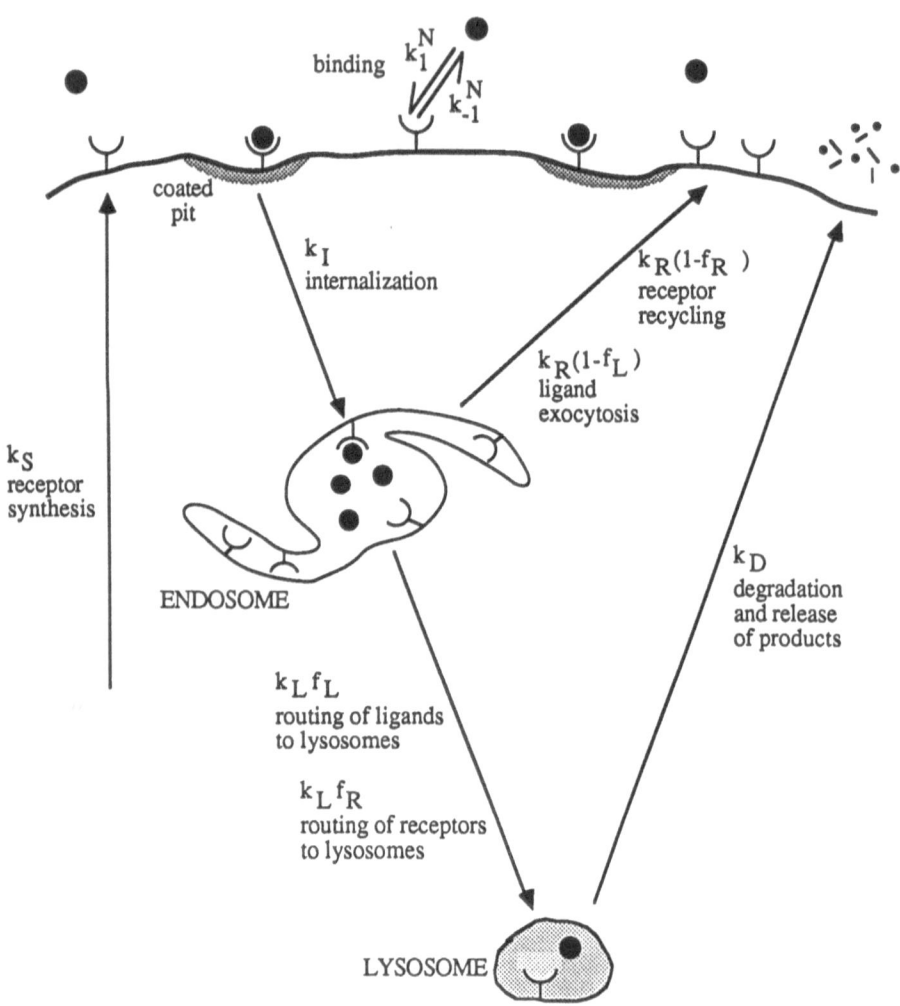

binding k_1^N

k_{-1}^N

coated pit

k_I
internalization

$k_R(1-f_R)$
receptor
recycling

$k_R(1-f_L)$
ligand
exocytosis

k_S
receptor
synthesis

k_D
degradation
and release
of products

ENDOSOME

$k_L f_L$
routing of ligands
to lysosomes

$k_L f_R$
routing of receptors
to lysosomes

LYSOSOME

FIGURE 5.1 Whole cell kinetic model of the endocytic cycle. Rate constants are assigned to the steps in the cycle corresponding to binding and dissociation at the cell surface, internalization, recycling, routing of molecules to the lysosome, and degradation. New receptor synthesis occurs at a constant rate. The outcome of the sorting process is reflected in the values of the sorting fractions f_R, the fraction of receptors degraded, and f_L, the fraction of ligand molecules degraded.

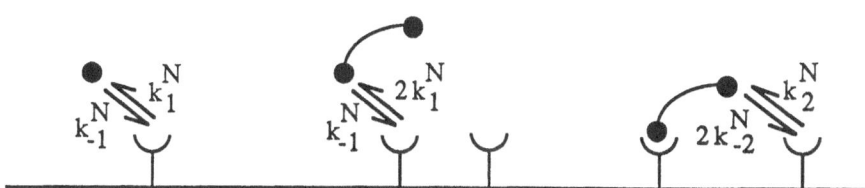

FIGURE 5.2   Binding, dissociation, and crosslinking at the cell surface. Characterization of the binding of multivalent ligands to the cell surface is more complex than that of monovalent ligands. Bivalent ligands with two identical binding sites, for example, may crosslink two surface receptors. The rate constants describing the interaction of the second binding site on the ligand with a second receptor are in general not equal to the rate constants describing the interaction of the first binding site on the ligand with the first receptor. In addition, statistical factors must be included when there is more than one identical reaction which may occur with equal probability.

k_D is the rate constant for the degradation of molecules in lysosomes and the transport of fragments to the cell surface; k_S is the rate of new receptor synthesis.

We also include terms in the whole cell kinetic model to characterize the sorting step. Tubules remove contents from the endosome, and both the rate of movement of tubules back to the cell surface and the contents of the tubules are important. We have chosen k_R to represent the intrinsic rate constant for the vesicular transport of material from the endosome back to the cell surface. This rate constant is a function of both the sorting time, defined as the time the tubule and vesicle are connected so that the tubule can receive molecules from the vesicle, and the rate of movement of the tubule and its contents through the cytoplasm to the cell surface. We assume that k_R does not depend on the endosome geometry or on ligand or receptor properties. The contents of the tubule are denoted by $(1-f_R)$, the fraction of endocytosed receptors contained in the tubules, or $(1-f_L)$, the fraction of endocytosed ligand in the tubules. The sorting process itself determines these fractions. If the diffusion with trapping mechanism we developed to explain sorting in Chapter 4 is correct, then these fractions depend only on the sorting time, the receptor diffusion coefficient, the geometry of the endosome, the dimensionless membrane current, and the ligand size.

Similarly, the routing of molecules to the lysosome is described by the product of k_L, the rate constant representing vesicular transport of the remaining endosome material to lysosomes, and the fraction of endocytosed receptors, f_R, or endocytosed ligand, f_L, left behind by the tubules. The vesicular rate constant k_L is a function of the sorting time and the time it takes for the vesicle to find and fuse with a lysosome or to mature into a lysosome and, like k_R, is assumed to be independent of the endosome geometry and of the ligand and receptor identities.

More complex systems, although not considered here, might also be might analyzed. Our general scheme assumes that all rate constants are independent of time; some systems may in fact require that one or more of the rate constants are time dependent. As mentioned in Chapter 2, for example, K562 cells were shown to alter their rate of synthesis of transferrin receptors in accordance with their success in internalizing diferric transferrin (Weissman et al. 1986). The mannose-6-phosphate receptor system is a complex system for another reason: the receptor is known to bind both endogenous ligands present in the Golgi and exogenous ligands present in the medium. Because the receptor is believed to cycle between the Golgi, endosomes, and the cell surface, the analysis of this system would require detailed information about the route of receptor travel (Gonzalez-Noriega et al. 1980; Brown et al. 1986).

5.3 Model Equations

The kinetic equations describing the binding of ligand to cell surface receptors, internalization, and the recycling or degradation of molecules can be written quite simply using the notation of Figure 5.1. A more complicated situation arises if the ligand and possibly also the receptor are multivalent, for all of the relevant binding and crosslinking steps must be included. In addition, there is evidence in several receptor systems that free receptors can also be internalized (Anderson et al. 1982; Watts 1985), and this possibility must also be reflected in the equations. It is difficult to solve these nonlinear equations and to fit all of the model parameters simultaneously to the type of experimental whole cell kinetic data that are presently available. We therefore make an important assumption.

In many instances, internalization experiments are carried out under "saturating" ligand concentrations, implying, at least for monovalent ligands and receptors, that the ligand concentration is much greater than the equilibrium dissociation constant, k_{-1}^N/k_1^N, and that therefore all of the surface receptors can be assumed to be occupied at equilibrium. Experiments with multivalent ligands are also typically performed at high ligand concentrations so that at equilibrium nearly all of the receptors can be assumed to be occupied. Binding equilibrium may occur rapidly in many systems, as predicted from association and dissociation rate constant data (e.g. Table 6.1). In addition, surface binding is often allowed to equilibrate prior to the start of an experiment by incubating cells for 1 to 2 hours at 4 °C, a temperature at which no internalization occurs. Thus we make the assumption that the binding of ligand to cell surface receptors is at equilibrium and that the concentration of ligand is great enough that virtually all of the surface receptors are bound. These assumptions insure that the nonlinear terms in the original full description of the endocytic cycle, the binding and crosslinking terms, are no longer present.

We define R_s, R_i, R_l, and R_d as the dimensionless numbers of bound surface receptors, receptors in the endosome, receptors in the lysosome, and degraded receptors, respectively. Each is equal to the total number of receptors in that state or location divided by the total number of surface receptors at time t_0, R_0. The equations for receptor distribution as a function of time can then be written:

$$\frac{dR_s}{dt} = -k_I R_s + k_R (1-f_R) R_i + k_S \tag{5.1}$$

$$\frac{dR_i}{dt} = k_I R_s - k_R (1-f_R) R_i - k_L f_R R_i \tag{5.2}$$

$$\frac{dR_l}{dt} = k_L f_R R_i - k_D R_l \tag{5.3}$$

$$\frac{dR_d}{dt} = k_D R_l. \tag{5.4}$$

The release of fragments of the degraded receptors to the medium is rapid and therefore we do not need to take into account degraded receptors that have not yet diffused out of the cell into the medium.

The distribution and accumulation of ligand molecules in the cell can be described similarly. L_s, L_i, L_l, and L_d represent the dimensionless number of ligand molecules present on the cell surface, in endosomes, in lysosomes, and that have been degraded. Each is made dimensionless by dividing the number of ligand molecules in that state by the number of surface receptors at time t_0, R_0. Thus the relevant equations are

$$\frac{dL_i}{dt} = k_I^L L_s + \frac{N_A V L_0}{R_0} - k_R (1-f_L) L_i - k_L f_L L_i \qquad (5.5)$$

$$\frac{dL_l}{dt} = k_L f_L L_i - k_D^L L_l \qquad (5.6)$$

$$\frac{dL_d}{dt} = k_D^L L_l, \qquad (5.7)$$

where N_A is Avogadro's number, V is the total volume of endocytic vesicles internalized per unit time, and L_0 is the concentration of ligand in the medium. The term $N_A V L_0 / R_0$ represents the nonspecific uptake of ligand molecules via the fluid that fills every forming endocytic vesicle. In the most general case, the internalization rate constant k_I^L may be different from the internalization rate constant for the receptor, k_I. The ligand degradation rate constant k_D^L may be different from the receptor degradation rate constant k_D because of differences in the structure of the two molecules and therefore in the action of the lysosomal enzymes in breaking down that structure. Note that the assumption that all surface receptors are bound is implicit in the use of these equations; alternatively, equations could be written to include the binding step in the cycle.

A relationship between R_s and L_s remains to be specified. Because we have assumed that all surface receptors are bound due to the presence of a saturating concentration of ligand, if both the receptor and ligand are monovalent, $R_s = L_s$. If instead the ligand is multivalent, a single ligand molecule may bind to several receptors at once, crosslinking them. In this case, $R_s \geq L_s$. If we assume that the binding of the multivalent ligand to the monovalent receptors quickly reaches equilibrium, we then need only to know the equilibrium number of ligand molecules bound to one, two, three, etc. receptor molecules simultaneously. For the case of monovalent receptors and bivalent ligands, for example, Perelson and DeLisi (1980) have calculated the fraction of receptors bound but not crosslinked, f_b, and the fraction of receptors crosslinked, f_c, by the ligand as a function of the equilibrium binding and crosslinking constants, ligand concentration, and surface receptor density. The desired relationship between the number of surface receptors and the amount of surface-bound ligand is then simply

$$L_s = R_s (f_b + \frac{f_c}{2}).$$ (5.8)

Similar relationships can be found for the case of monovalent receptors bound by ligands of greater valency. A more difficult situation to analyze is that of multivalent receptors bound by multivalent ligands; Macken and Perelson (1985) have used the mathematics of branching processes to describe the distribution of aggregate sizes when f-valent ligands bind and crosslink g-valent receptors (f and g $\geq$ 1).

The relationship between the receptor and ligand internalization rate constants, k_I and k_I^L, is related to the relationship for R_S and L_S and may be further complicated if, for example, crosslinked receptors are internalized at a greater rate than bound receptors that are not crosslinked.

Regardless of the relationship specified between R_S and L_S, the time-dependent solution to the linearly independent equations describing receptor movement through the cell, Eqs. 5.1 - 5.3, can be found. Linear algebra techniques were used to identify the state transition matrix $\underline{\underline{\Phi}}(t,t_0)$ as:

$$\Phi(t,t_0) = \begin{bmatrix} ae^{m_2 (t-t_0)} - be^{m_3 (t-t_0)} & c(e^{m_3 (t-t_0)} - e^{m_2 (t-t_0)}) & 0 \\ d(e^{m_2 (t-t_0)} - e^{m_3 (t-t_0)}) & ae^{m_3 (t-t_0)} - be^{m_2 (t-t_0)} & 0 \\ ge^{-k_d (t-t_0)} (he^{m_2 (t-t_0)} - ie^{m_3 (t-t_0)}) & je^{-k_d (t-t_0)} - le^{m_2 (t-t_0)} + me^{m_3 (t-t_0)} & e^{-k_d (t-t_0)} \end{bmatrix}$$ (5.9)

where

$$\alpha_1 = k_I + k_R(1-f_R) + k_L f_R$$ (5.10)

$$\alpha_2 = k_I k_L f_R$$ (5.11)

$$m_2 = \frac{-\alpha_1 + \sqrt{\alpha_1^2 - 4\alpha_2}}{2}$$ (5.12)

$$m_3 = \frac{-\alpha_1 - \sqrt{\alpha_1^2 - 4\alpha_2}}{2} \tag{5.13}$$

$$a = \frac{k_R(1-f_R) + k_L f_R + m_2}{m_2 - m_3} \tag{5.14}$$

$$b = \frac{k_R(1-f_R) + k_L f_R + m_3}{m_2 - m_3} \tag{5.15}$$

$$c = \frac{ab \, (m_2 - m_3)}{k_I} \tag{5.16}$$

$$d = \frac{k_I}{m_2 - m_3} \tag{5.17}$$

$$g = \frac{k_I k_L f_R}{(k_D + m_2)(k_D + m_3)} \tag{5.18}$$

$$h = \frac{k_I k_L f_R}{(m_2 - m_3)(k_D + m_2)} \tag{5.19}$$

$$i = \frac{k_I k_L f_R}{(m_2 - m_3)(k_D + m_3)} \tag{5.20}$$

$$j = k_L f_R \left[\frac{d_2}{k_D + m_2} - \frac{d_1}{k_D + m_3} \right] \tag{5.21}$$

$$l = \frac{k_L f_R d_2}{k_D + m_2} \tag{5.22}$$

$$m = \frac{k_L f_R d_1}{k_D + m_3} \, . \tag{5.23}$$

The solution depends on this fundamental matrix $\underline{\underline{\Phi}} \, (t, t_0)$ and the choice of initial conditions according to

$$\begin{bmatrix} R_s(t) \\ R_i(t) \\ R_l(t) \end{bmatrix} = \underline{\underline{\Phi}} \, (t, t_0) \begin{bmatrix} R_s^0 \\ R_i^0 \\ R_l^0 \end{bmatrix} + \int_{t_0}^{t} \underline{\underline{\Phi}} \, (t, s) b(s) \, ds, \tag{5.24}$$

where $[\ R_s^0,\ R_i^0,\ R_l^0\ ]^T$ is the vector of initial conditions and $b(s)$ is the vector $[\ k_S,\ 0,\ 0\ ]^T$. If a relationship between R_S and L_S is specified, the time-dependent solution of Eqs. 5.5 - 5.7 can also be found.

One particular case of the above deserves special consideration. A type of experiment in which the fates of receptors can be followed directly is one in which the receptors themselves are labeled. Ukkonen et al. (1986) iodinate the surface Fc receptors of macrophages at time 0 and then follow the movement of the labeled receptors through the cell as a function of time. Thus the initial condition for this experiment is that $R_s^0 = 1$ and $R_i^0 = R_l^0 = R_d^0 = 0$: the vector of initial conditions is known exactly. In addition, the rate of synthesis of new receptors, k_S, can be set to zero because no new labeled receptors can appear after the initial labeling is completed. For these labeled receptor experiments, then, the solution for R_s, R_i, R_l, R_d as a function of time is:

$$R_s(t)\ =\ \frac{1}{m_2-m_3}\left[\ e^{m_2 t}(k_R+k_L f_R+m_2) - e^{m_3 t}(k_R+k_L f_R+m_3)\ \right] \tag{5.25}$$

$$R_i(t)\ =\ \frac{k_I}{m_2-m_3}\left[\ e^{m_2 t} - e^{m_3 t}\ \right] \tag{5.26}$$

$$R_l(t)\ =\ \frac{k_I k_L f_R}{m_2-m_3}\left[\ \frac{e^{m_2 t}}{k_D+m_2} - \frac{e^{m_3 t}}{k_D+m_3} + \frac{m_2-m_3}{(k_D+m_2)(k_D+m_3)}\ e^{-k_D t}\ \right]. \tag{5.27}$$

In addition,

$$R_d(t)\ =\ 1 - (\ R_s(t) + R_i(t) + R_l(t)\ ). \tag{5.28}$$

Note that there are fewer unknowns in the solution expressed in Eqs. 5.25 - 5.28 than in the more general formulation of the solution expressed in Eq. 5.24. Therefore, in attempting to use our whole cell model to fit experimental data away from steady state and thus identify many of the rate constants simultaneously, a recommended first step is to label surface receptors at time 0 and to follow the movement of those receptors as a function of time. Once these data are obtained and the relevant rate constants are fit, additional experiments may be performed in which the synthesis rate k_S is measured and the ligand movement through the cell is monitored.

5.4 Use of the model

Obtaining whole cell kinetic data

Whole cell kinetic data on the endocytic cycle have been obtained for a number of receptor/ligand systems using a variety of experimental methods. A typical experiment performed in order to follow the kinetics of ligand internalization is as follows (Mellman and Plutner 1984; Zigmond et al. 1982; Schwartz et al. 1982; Hopkins and Trowbridge 1983). Ligand molecules are labeled with ^{125}I or ^{3}H and allowed to bind to surface receptors at 37 ^{o}C. Alternatively, cells may be preincubated with labeled ligand at 4 ^{o}C, to allow binding but no internalization, and then warmed to 37 ^{o}C. At various times after internalization has begun, the cells are chilled to 4 ^{o}C to prevent further internalization, washed to remove unbound ligand, and analyzed. The total amount of cell-associated ligand is measured, giving the total of L_s, L_i, and L_l. Acid or other treatments are used to rapidly strip bound ligand from the cell surface, the only ligand accessible to the medium. The radioactivity associated with the cell is again measured, giving the total of L_i and L_l. Thus one can obtain $L_s(t)$ and $L_i(t) + L_l(t)$ from this type of experiment. Cell fractionation can further give the distribution of ligand inside the cell, although this is not commonly done. For example, lysosomes can be isolated by fractionating cells and centrifuging in a Percoll density gradient; the fractions containing acid phosphatases contain the lysosomes (Alberts et al. 1983; Ukkonen et al. 1986). In addition, a technique has recently been developed by which endosomes may be isolated from other cell membranes and organelles (Marsh et al. 1987), and this technique may soon enable the determination of $L_i(t)$. The accumulation of degraded ligand appearing in the medium can also be monitored to obtain $L_d(t)$.

The above experiment is also performed with one significant variation (Mellman et al. 1984; Marshall 1985a; Townsend et al. 1984). Cells are allowed to bind and internalize labeled ligand at 37 ^{o}C in order to load endosomes with endocytosed ligand. The cells are then washed to remove unbound ligand, treated at 4 ^{o}C with acid or another agent to remove any surface bound ligand, and placed in medium free of ligand at 37 ^{o}C. The degradation of ligand can then be monitored as in the earlier experimental technique. In addition, however, the exocytosis of free ligand can be detected by the appearance of undegraded ligand in the medium. The appearance of bound ligand at the cell surface, if detected, can be assumed to represent the exocytosis of ligand remaining bound to receptors because the concentration of undegraded ligand in the medium is usually very small.

These experiments, and similar ones, are commonly used to follow the kinetics of ligand internalization and degradation. A typical result of such experiments is shown in Figure 5.3.

The kinetics of receptor internalization, recycling, and degradation are reported with less

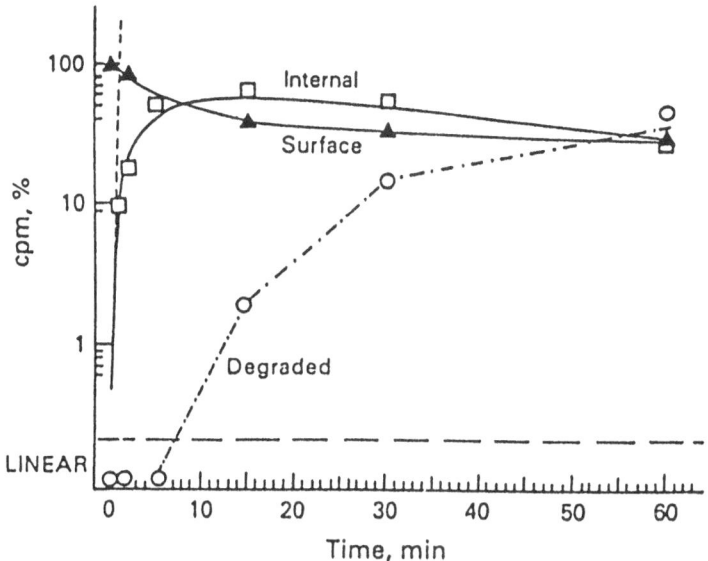

FIGURE 5.3 Typical experimental data on the kinetics of ligand internalization and degradation. The data are taken from Bridges et al. (1982) and show the amount of surface bound, intracellular, and degraded ligand (in counts per minute) as a function of time for the asialoglycoprotein receptor system on rat hepatocytes with the ligand asialoorosomucoid. In this experiment, the ligand was incubated with the cells at 4 $^{\circ}$C prior to the start of the experiment. The x axis gives the time elapsed after raising the temperature to 37 $^{\circ}$C.

frequency. If the receptors can be labeled either biosynthetically or at the cell surface, for example, by surface iodination, their movement through the cell can be monitored (Ukkonen et al. 1986; Weissman et al. 1986). Cell fractionation and analysis of the individual fractions can give an indication of the location of the labeled receptors as a function of time. The degradation of receptors can be detected by the appearance of the radioactive label in the medium. In addition, experiments can also be performed to measure the rate of new receptor synthesis, as described in Chapter 2.

Fitting the experimental data

Experimental data on the kinetics of the endocytic cycle rarely supply the desired rate constants from Figure 5.1 directly. The apparent steady state rate of ligand internalization, for example, is the result of true internalization less any exocytosis of undegraded ligand. Thus the whole cell kinetic data must be analyzed with a whole cell model such as the one presented here in order to obtain the values of the rate constants and the values of the sorting fractions. This is the information needed in order to determine whether the effects of ligand on the dynamics of the endocytic cycle are a result of variations in the rate constants of steps in the endocytic cycle not related to sorting and/or variation in the outcome of the sorting process itself.

The analysis is performed by fitting experimental data to the whole cell model using an appropriate optimization algorithm. The values of the rate constants for each step can be identified, and information on the sorting process is therefore obtained. For example, the movement of receptors and ligands through the cell can be analyzed and the values of the recycling rate constants $k_R(1-f_R)$ and $k_R(1-f_L)$ can be obtained. The ratio of these two rate constants will give the ratio R_1 of receptors recycled to ligand recycled, $(1-f_R)/(1-f_L)$. Similarly, the ratio R_2 of the rate constants $k_R f_R$ and $k_R f_L$ can be calculated to give f_R/f_L. The sorting fractions f_R and f_L can then be found from

$$f_R = \frac{R_2(1-R_1)}{R_2 - R_1} \tag{5.29}$$

$$f_L = \frac{1 - R_1}{R_2 - R_1} . \tag{5.30}$$

Alternatively, the analysis of whole cell endocytic data on two ligands A and B which bind to the same receptor will give information on the ratio of receptors or ligand recycled for ligand A to receptors or ligand recycled for ligand B, $(1-f_R)^A/(1-f_R)^B$ or $(1-f_L)^A/(1-f_L)^B$.

The ratio $f_R{}^A/f_R{}^B$ or $f_L{}^A/f_L{}^B$ can also be calculated, and the individual sorting fractions found by expressions analogous to Eqs. 5.29 and 5.30.

Thus, although techniques have not yet been developed through which the sorting fractions f_R and f_L themselves can be directly measured, this information can be made available through an analysis of whole cell kinetic data. These sorting fractions calculated from experimental data can then be compared with the fractions we predict by a theoretical analysis of a particular sorting mechanism.

One optimization algorithm which can be used to fit the data is Powell's algorithm (Powell 1977). This algorithm employs iterative quadratic programming to select the values of a number of parameters that will minimize the value of a user-specified objective function. In this context, the parameters would include the desired rate constants. A typical objective function might be the sum of the squares of the differences between a experimentally measured quantity, for example, degraded receptors, and the predicted value of that quantity given the model equations and the values of the rate constants. Powell's algorithm allows the specification of both equality and inequality constraints in the minimization of the objective function. For example, one can require that the rate constants are greater than zero and that the optimized values of the parameters must predict the values of certain quantities, such as degraded receptors, to within 20% of their experimentally measured values. This last example is especially useful for biological data, which typically exhibit a great deal of variance.

5.5 Application of the model to the Fc_γ receptor system

Experimental data

Mellman and coworkers (Mellman et al. 1983; Mellman et al. 1984; Mellman and Plutner 1984; Ukkonen et al. 1986) have studied the kinetics of the endocytic cycle for the Fc_γ receptor system on the mouse macrophage cell line J774. This receptor binds to the Fc portion of immunoglobulin G (IgG) molecules and functions to bind complexes of IgG molecules and antigen that have been produced by the action of the immune system and to internalize and degrade these complexes in order to eliminate them from the body. The particular Fc_γ receptor studied is the trypsin-resistant receptor for IgG1/IgG2b- containing immune complexes, in this chapter refered to as the Fc receptor. Mellman and coworkers investigated the kinetics of the endocytic cycle for two different ligands to the Fc receptor:

(1) the monovalent Fab fragment of the bivalent antibody 2.4G2, a high affinity monoclonal antibody to the Fc receptor, and

(2) multivalent IgG complexes formed by incubation of rabbit anti-dinitrophenyl (DNP)

IgG with DNP-modified bovine serum albumin (BSA).

The kinetics of the endocytic cycle were found to differ in several aspects, including receptor degradation and ligand exocytosis, between cells treated with the two ligands. One general finding was that more Fc receptors were degraded in the presence of the multivalent ligand than in the presence of the monovalant ligand. This enhanced degradation could be the result of at least two factors: receptors could be less efficiently returned to the cell surface in the presence of the multivalent ligand than in the presence of the monovalent ligand due to a decrease in the fraction of receptors recycled, or the sorting outcome could be identical in the two cases but the multivalent ligand could enhance the internalization rate. In this latter case, the enhanced internalization rate would mean that the multivalent ligand sends receptors through an imperfect sorting process more often and thus more receptors would be slated for degradation than in the monovalent ligand case. In order to examine whether the difference in cycle kinetics between the ligands was due to a difference in the outcome of the sorting process and/or to some other step in the cycle, we analyzed the Fc receptor data of Mellman and coworkers with our whole cell model.

The set of linear equations 5.1 - 5.7 describing the movement of receptors through the cell are valid if all surface receptors can be assumed occupied. We must first determine whether this assumption is approximately valid for the experiments of Mellman and coworkers on the Fc receptor systems. Experiments with the monovalent ligand were carried out at ligand concentrations of 1 to 5 μg/ml. The molecular weight of an Fab fragment falls in the range of 45,000 to 50,000; thus the concentration of Fab fragments was no less than 2×10^{-8} M. This is substantially greater than the concentration required for half-maximal binding, for the equilibrium dissociation constant K_D for this ligand is less than 10^{-9} M (Mellman et al. 1984). Because most of the experiments included a prebinding step and because this ligand concentration is great enough to effectively saturate all surface receptors at equilibrium, we assume that the surface receptors are occupied.

The analysis is more difficult in the case of the multivalent ligand. The total concentration of IgG in the medium when ligand (2) was used in experiments was 20 μg/ml. This concentration was found empirically to saturate all surface receptors; a binding experiment was performed in which the amount of IgG bound to surface receptors was found to plateau at 20 μg/ml (Mellman and Plutner 1984). Because essentially all surface receptors can be assumed to be bound by ligand, the linear Eqs. 5.1 - 5.7 describe the system.

Several different types of experimental data were obtained by Mellman and coworkers. The data listed below give information on the movement of receptors through the cells during the endocytic cycle and therefore the solution to model Eqs. 5.1 - 5.4 will be used to fit the data. The data used are:

(1) Rate of degradation of receptors. Receptors on the cell surface at time 0 were radioiodinated at 4 °C using lactoperoxidase and glucose oxidase, producing ^{125}I-Fc receptors. Cells were then incubated at 37 °C in the presence of saturating concentrations of ligand. The fraction of labeled receptors remaining undegraded was measured at times ranging from 0 to 6 hours after the start of the incubation. The difference between the initial number of labeled receptors and the number remaining at time t gives the fraction of receptors degraded, or $R_d(t)$. It was found that for ligand (2) R_d(120 min.) ≈ 0.28 and R_d(240 min.) ≈ 0.46. Cells exposed to ligand (1) acted as did control cells: R_d(120 min.) ≈ 0.08 and R_d(240 min.) ≈ 0.15. Ligand (1)-treated and control cells exhibited a receptor half-life of about 15 hours (Mellman and Plutner 1984; Mellman, personal communication).

(2) Fraction of receptors in the lysosome. Receptors on the cell surface at time 0 were labeled with ^{125}I. Cells were then incubated at 37 °C in the presence of saturating concentrations of ligand for 0 to 2 hours. Cells were lysed and fractionated and the fraction of undegraded receptor in the lysosomes was determined. For ligand (1), $R_l/(R_s+R_i+R_l) = 0.04$ at 60 min. and 0.132 at 120 min. For ligand (2), $R_l/(R_s+R_i+R_l) = 0.248$ at 60 min. and 0.270 at 120 min. (Ukkonen et al. 1986).

(3) Apparent internalization rate constant for monovalent and multivalent ligands. Labeled ligand was bound to cell surface at 4 °C. Cells were then warmed to 37 °C and the rate of internalization of the label was obtained by measuring the rate at which the label became resistant to removal by low pH treatment or the protease subtilisin, which remove ligand bound to surface receptors. The half-time for ligand internalization was found to be on the order of 15 to 20 min. for the monovalent Fab ligand (Mellman et al. 1984) and 1 to 2 min. for the multivalent IgG complexes (Mellman and Plutner 1984). Because the ligands are internalized while bound to receptors, these ligand half-times for internalization correspond to internalization rate constants k_i on the order of 0.04 to 0.05 min^{-1} for receptors bound by the monovalent ligand and 0.35 to 0.69 min^{-1} for receptors bound by the multivalent ligand.

These apparent rate constants are underestimates of the true internalization rate constants because they are uncorrected for any receptor recycling which may be occurring. The measured half-time for the monovalent ligand experiment is particularly unlikely to give an accurate estimate of the true internalization rate constant. Mellman et al. (1984) found that 45 to 73% of internalized ^{125}I-Fab reappeared at the cell surface, still bound to receptors, within 10 minutes of incubation at 37 °C, corresponding to a recycling rate constant on the order of 0.2 min^{-1}. These data suggest that much of the internalized ligand is exocytosed, possibly in a receptor-bound state. Thus there is a substantial amount of recycling occurring, and the apparent internalization rate constant of 0.04 to 0.05 min^{-1} is likely a substantial underestimate of the true internalization rate constant when receptors are bound by the monovalent ligand.

(4) Minimum estimate of the internalization rate constant. The membrane molecules of a cell are constantly being internalized and recycled due to the constant vesicle traffic inside a cell. It was found that macrophages or J774 cells internalize at least 2 times their surface area per hour, and, by one measurement, as much as 23% in 5 minutes (Mellman et al. 1980; Steinman et al. 1976). Therefore, the internalization rate for a membrane molecule that is neither actively included or excluded by this process is on the order of 0.04 to 0.05 min^{-1}. If the Fc receptors are not hindered in their ability to reach an internalization site, presumably a coated pit, the internalization rate constant k_i may be assumed to have a minimum value of 0.04 min^{-1}. If receptors are actively included in the formation of internal vesicles, as might be expected if the receptors are trapped with some efficiency in coated pits, the internalization rate constant may be much greater than this value. This is most likely the case here, given the internalization data in (3) above.

Fitting the experimental data

The above experimental data were analyzed with our whole cell model in order to determine the values of the rate constants describing the movement of Fc receptors through the endocytic cycle when the cells are exposed to either the monovalent ligand Fab or the multivalent IgG complexes. Because only surface receptors were labeled in experiments (1) and (2), the initial condition for the experiments to be described is simply $R_s^0 = 1$ and $R_i^0 = R_l^0 = R_d^0 = 0$. Thus the results of these experiments may be described by the time-dependent solutions to Eqs. 5.1 - 5.4 given in Eqs. 5.25 - 5.28.

The data fit was performed in the following way. The four unknown rate constants, k_I, k_R $(1-f_R)$, $k_L f_R$, and k_D, were varied by both trial and error methods and within the context of an optimization routine, Powell's algorithm (Powell 1977), in order to minimize the value of an objective function which measured the ability of the fit parameter values to predict the experimental data. The objective function OBJFUN was defined by:

$$\text{OBJFUN} = (R_s^{theor}(t_1) - 0.5)^2 + (R_d^{theor}(t_3) - R_d^{exp}(t_3))^2 + (R_d^{theor}(t_4) - R_d^{exp}(t_4))^2$$

$$+ \left[\left[\frac{R_1}{R_s + R_i + R_1} \right]^{theor}(t_2) - \left[\frac{R_1}{R_s + R_i + R_1} \right]^{exp}(t_2) \right]^2$$

$$+ \left[\left[\frac{R_1}{R_s + R_i + R_1} \right]^{theor}(t_3) - \left[\frac{R_1}{R_s + R_i + R_1} \right]^{exp}(t_3) \right]^2, \tag{5.31}$$

where the superscripts theor and exp refer to the calculated and experimental values of the indicated quantities. The time t_1 is the observed half-time for internalization (15 to 20 min. for the monovalent ligand and 1 to 2 min. for the multivalent ligand), $t_2 = 60$ min., $t_3 = 120$ min., and $t_4 = 240$ min. All rate constants were constrained to take on values greater than or equal to zero.

The results of the data fit are given in Table 5.1 and plotted in Figure 5.4. For both ligands, it was possible to identify values for the rate constants for which the value of the objective function was less than or equal to 0.010. For reference, note that the value of the objective function would be 0.0125 if each of the five calculated values in the objective function differed from the experimental values by only 5%.

Because the rate constant for removing receptors from the endosome to the lysosome was found to be much less than the rate constants for internalization and recycling and because the data available for the fitting did not give information on the receptor cycle time, the initial rate of internalization, or the number of receptors in the endosome as a function of time, it was not possible to uniquely identify the values of the rate constants for internalization and recycling, k_I and k_R $(1-f_R)$. However, it was possible to identify the ratio of these two rate constants k_R $(1-f_R)/ k_I$, in addition to the minimum (apparent) value of the rate constant k_I derived earlier.

The rate constants for degradation were found to be similar for the two ligands, and the differences are presumably due to differences in the action of the lysosomal proteases on the two ligand structures.

The rate constants for the routing of receptors from endosomes to lysosomes were found to be significantly different for the two types of ligand. For the monovalent ligand, $k_L f_R$ is approximately equal to 0.003 min^{-1}; for the multivalent ligand, $k_L f_R$ is approximately equal to 0.008 min^{-1}. We assume that the rate constant k_L is a function of the sorting time and the time it takes for the endosomal vesicle to find and fuse with a lysosome or to mature into a lysosome and that this rate constant is independent of the ligand identity. Thus the difference in the overall rate constant $k_L f_R$ between the two ligands is a result of a difference in the fraction of receptors routed to the lysosome, or f_R. Our analysis of the Fc receptor data suggests that the

ligand	k_I^{min} (min^{-1})	$\dfrac{k_R(1-f_R)}{k_I}$	$k_L f_R$ (min^{-1})	k_D (min^{-1})	OBJFUN
(1) Fab fragments	0.04 - 0.05	~ 1.0	0.003	0.007	0.002
(2) IgG complexes	0.35 - 0.69	~ 0.5	0.008	0.011	0.010

TABLE 5.1 Results of fitting the receptor data of Mellman and co-workers with the whole cell kinetic model. The minimum value of the internalization rate constant k_I^{min}, equal to the apparent internalization rate constant, the ratio of the recycling to internalization rate constant $k_R (1-f_R)/k_I$, the rate constant for routing receptors to the lysosome $k_L f_R$, and the degradation rate constant k_D are given. The value of the objective function, an indication of the goodness of fit to the experimental data, is also listed.

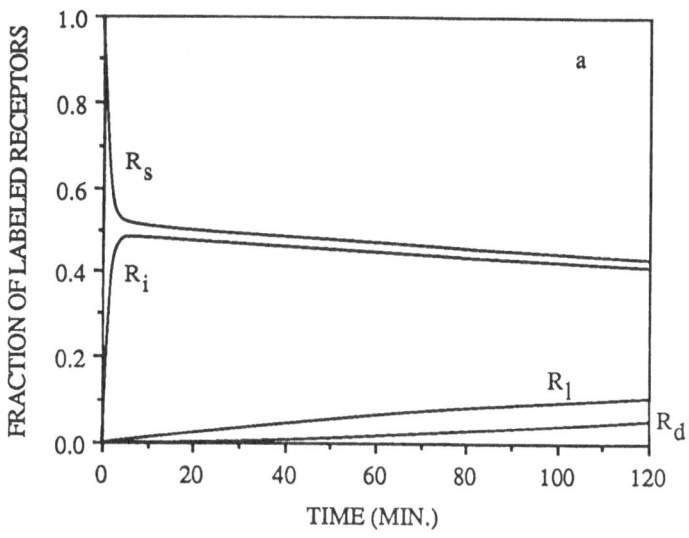

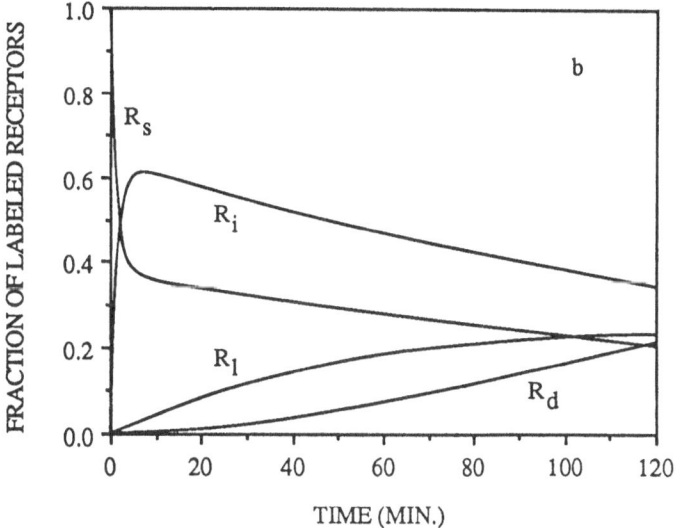

FIGURE 5.4 Sample fits to the data of Mellman and co-workers. Although only the ratio $k_R(1-f_R)/k_I$ can be identified by the data fit, values for k_I and k_R $(1-f_R)$ were chosen here in order to plot typical fits to the data. a) Monovalent data fit. Parameter values: $k_I = 0.48$, $k_R(1-f_R) = 0.50$, $k_L f_R = 0.003$, $k_D = 0.007$. b) Multivalent data fit. Parameter values: $k_I = 0.48$, $k_R(1-f_R) = 0.28$, $k_L f_R = 0.008$, $k_D = 0.011$.

fraction of receptors sent to lysosomes from the sorting process when the cell is exposed to the multivalent ligand is nearly three times the fraction sent to lysosomes when the cell is exposed to the monovalent ligand. Thus we have shown that the differences in the kinetics of the endocytic cycle observed by Mellman and coworkers upon changing the identity of the ligand are due at least in part to differences in the outcome of the sorting process itself.

Finally, we can suggest the data needed to uniquely identify the values of the receptor internalization and recycling rate constants in each of the two cases and thus the ratio $(1-f_R)^A/(1-f_R)^B$, where A and B denote the monovalent and multivalent ligands. Techniques have recently been developed for the isolation of endosomes (Marsh et al. 1987), making possible the determination of $R_i(t)$ in the labeled receptor experiments. Alternatively, a measurement of the initial rate of receptor internalization, taken before recycling is expected to begin, could be obtained. This type of data, together with the data obtained by Mellman and coworkers and listed earlier, would enable a confident identification of the relevant rate constants. The full set of rate constants could then be analyzed by a method analogous to that given in Eqs. 5.29 and 5.30 to give f_R^A and f_R^B.

Possible explanations for the difference in the sorting outcome

The fitting of our whole cell model to experimental data on the Fc receptor system in macrophages shows that the outcome of sorting may be affected by the ligand identity. It is not yet clear what properties of the ligand may be responsible for the different outcome, but some observations may be made at this point. Mellman and coworkers (Mellman et al. 1984; Mellman and Plutner 1984) speculated that the difference in ligand valency between these two ligands might result in the difference in the sorting outcome. However, they realized that the two ligands bind to different portions of the Fc receptor and that this difference, through some unknown mechanism, might also be the cause of the differences observed when cells were treated with these two ligands. Therefore, multivalent complexes of the Fab ligand (1) were formed by the adsorption of monovalent Fab fragments onto 5 to 10 nm. colloidal gold particles. Experiment (2) was repeated with this new ligand with essentially the same results as if IgG complexes were used: $R_l/(R_s+R_i+R_l) = 0.256$ at 60 min. and 0.312 at 120 min. (Ukkonen et al. 1986). This suggests that the difference in processing of the monovalent ligand (1) and the multivalent ligand (3) is likely due to a difference in a ligand property, probably valency, and not to a difference in the ligand binding site on the receptor.

In addition, we note that the ligand/receptor complexes of the monovalent and multivalent complexes may be affected differently by the low pH of the endosome. The monovalent Fab fragments are likely to dissociate from the Fc receptors at low pH (Mellman et al. 1984) but the IgG complexes are relatively insensitive to low pH (Mellman and Plutner 1984).

5.6 Discussion

In this chapter, we developed a whole cell model of the endocytic cycle in order to obtain quantitative information on the outcome of the sorting process from the only type of data available at this time, whole cell data on the kinetics of the endocytic cycle. For example, data on the changes in the number of cell surface receptors with time upon exposure of cells to different ligands, data on the fraction of receptors and cell-associated ligand that are located intracellularly at various locations and specific times, and data on the rate of new receptor synthesis, receptor degradation, and ligand degradation can be analyzed by using this model and an appropriate optimization algorithm. In this way, information on the sorting fractions f_R and f_L can be obtained. These values, derived from analysis of experimental data, can then be compared with the theoretical sorting fractions predicted from a variety of single endosome sorting models. A general scheme for using this whole cell model in the context of analyzing the sorting process is shown in Figure 5.5.

The predictions of the diffusion with trapping sorting model proposed in Chapter 4 can now be compared quantitatively with experimental data. Although the mechanism met the basic requirements of sorting derived in Chapter 2, it fails to meet more rigorous requirements. The analysis of the Fc receptor data of Mellman and coworkers (Mellman et al. 1984; Mellman and Plutner 1984; Ukkonen et al. 1986), for example, shows that the fraction of receptors slated for degradation at the completion of sorting depends on the identity of the ligand. Our analysis of their data shows that fewer receptors are recycled in the case of a multivalent ligand, IgG complexes, than in the case of a monovalent ligand, Fab fragments of an anti-receptor antibody. This effect is not predicted by the diffusion with trapping model. In addition, Mellman and co-workers have observed that the monovalent ligand is exocytosed in amounts much greater than the multivalent ligand, suggesting that the fraction of ligand degraded, f_L, also varies substantially with the ligand identity. This variation is likely much greater than would be predicted by the diffusion with trapping model, which allows for variation in ligand exocytosis only as a result of a change in the partition coefficient.

A similar whole cell analysis has been performed on the data of Schaudies et al. (1985), who investigated the time-course of binding and internalization of 125_I-epidermal growth factor in responsive and nonresponsive cell lines. It was shown that one possible contributing factor to the very different levels of responsiveness in the different cells is a significant variation in the outcome of the sorting process (Lauffenburger et al. 1987). Such differences are not predicted by the diffusion with trapping model.

The diffusion with trapping model also fails to predict ligand exocytosis in amounts greater than about 30 to 40%, and this fraction has been shown to be exceeded both with the

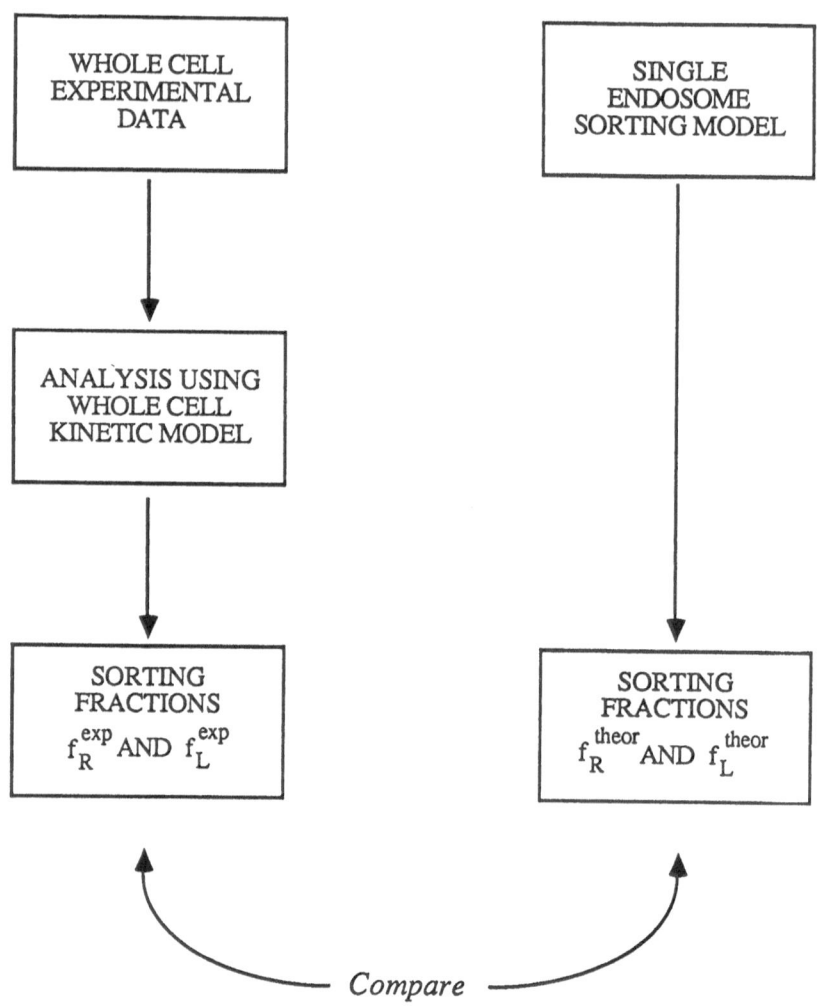

FIGURE 5.5 General scheme for analysis of the sorting process. Experimental whole cell experimental data are analyzed by the whole cell kinetic model presented in this chapter to give information on the "experimental" sorting fractions, f_R^{exp} and f_L^{exp}. The analysis of single endosome models of the sorting process, such as the diffusion with trapping model presented in Chapter 4, give predictions of the "theoretical" sorting fractions, f_R^{theor} and f_L^{theor}. A comparison of experimental and theoretical sorting fractions will indicate the ability of any proposed sorting model to explain experimental data.

binding of the monovalent Fab fragment to Fc receptors (Mellman et al. 1984) and the binding of some asialoglycoproteins to the asialoglycoprotein receptor (Townsend et al. 1984).

Thus the proposed diffusion with trapping sorting mechanism is not sufficient to account for a variety of outcomes of the sorting process. For this reason, we develop a more complete single endosome model of sorting in Chapter 6.

CHAPTER 6
A KINETIC MODEL OF SORTING THAT PREDICTS LIGAND EFFECTS

6.1   Introduction

The deficiencies of the diffusion with trapping sorting mechanism developed in Chapter 4 suggest the need for a more complete single endosome model of sorting. Although this mechanism is successful in explaining the kinetics and efficiency of receptor recycling in many systems, the mechanism allowed for variations in the sorting outcome only as a result of variations in the values of the vesicle radius R, the tubule radius b, the distribution of volume between the endosome vesicle and tubules, the diffusion coefficient of the receptor D_R, the dimensionless current β, the partition coefficient κ, and the sorting time τ^*.

In reality, the sorting process has a wide variety of outcomes, many of which do not appear to be under the influence of just these simple parameters. In this chapter, we review the literature data on the wide range of outcomes of the sorting process and on the changes in the sorting outcome found upon altering only the ligand properties in a particular receptor system, changes that we term ligand effects on sorting. We then develop a model of sorting to explain and predict these sorting outcomes.

The outcome of the sorting process has been shown to vary among different receptor systems. Marshall (1985b) found that insulin receptors on rat adipocytes are efficiently recycled to the cell surface and that even after 4 hours of continuous ligand uptake, during which the receptors cycle through the sorting process about once every 6 min., there is no significant loss of receptors to the degradative pathway. During this time, about 75% of the internalized ligand is degraded and the remaining 25% is exocytosed. Yet the outcome of sorting is very different in the transferrin receptor system. The ligand transferrin and its receptor are not separated and both apparently recycle to the cell surface, the ligand still bound to its receptor (Ciechanover et al. 1983). In contrast, the opposite seems to occur in the epidermal growth factor receptor system in many cell types. Here, the evidence indicates that neither the receptor nor its ligand recycle (Stoscheck and Carpenter 1984).

In addition to differences in sorting between receptor systems, experimental studies have indicated that the outcome of the sorting process for one particular receptor system may depend on properties of the ligand molecules themselves, for example, ligand valency. Mellman et al. (1984) have shown that when mouse macrophage Fc$_\gamma$ receptors are bound by monovalent Fab fragments of an anti-receptor antibody, the receptors are recycled efficiently and much of the ligand is exocytosed. However, binding of those same receptors with either of two multivalent ligands, the same Fab fragments complexed with gold or IgG-gold complexes, results in a decrease of the receptor half-life from 15 to 5 hours as more receptors are routed to the degradative pathway (Mellman and Plutner 1984; Ukkonen et al. 1986). These results have been explained in detail in Chapter 5 and were analyzed to show that at least part of the difference in receptor half-life is the result of a variation in the sorting process. In this system, then, the valency of the ligand appears to be important in determining the outcome of the sorting process.

Ligand characteristics, in particular the ligand valency, may also affect the sorting outcome in the transferrin receptor system. Hopkins and Trowbridge (1983) found that the monovalent ligand transferrin is internalized and then exocytosed by A431 cells and that the number of cell surface receptors does not change appreciably during several rounds of ligand uptake. In contrast, an internalized anti-transferrin receptor antibody, a bivalent ligand, is partially degraded and the number of cell surface receptors reduced significantly upon the uptake of this second ligand. Both of these ligands are internalized at the same rate, so the difference in processing apparently lies in the sorting process. Similarly, Weissman et al. (1986) showed that the anti-transferrin receptor antibody OKT9 is internalized by K562 cells and results in the routing of more receptors to the degradative pathway than the normal ligand, transferrin.

The recycling of the low density lipoprotein (LDL) receptor is also affected by ligand properties. Anderson et al. (1982) found that both LDL, which is monovalent, and a monoclonal antibody to the receptor, which is bivalent, allow receptor recycling. Thus ligand valency may not be the sole determinant of the sorting outcome; in this chapter, we will suggest a role for ligand affinity as well. Ligand properties have also been shown to effect the sorting outcome in the asialoglycoprotein receptor system (Townsend et al. 1984; Schwartz et al. 1986).

Finally, ligand effects on sorting have been demonstrated for the immunoglobulin E (IgE) receptor on rat basophilic leukemia cells, the tumor analog of basophils. In this system, IgE molecules bind via their Fc portion to IgE receptors, making the monovalent receptor now appear functionally bivalent. If a bivalent antigen with which the IgE molecules are able bind is then introduced, receptors can be crosslinked. This is shown schematically in Figure 6.1. Furuichi and co-workers (Furuichi et al. 1986) found that after internalization, crosslinked receptors and their ligands are not recycled to the cell surface but are instead degraded in lysosomes. Receptors not crosslinked due to the presence of either no antigen or only monovalent antigen are recycled together with their bound ligand. Because the rate of internalization did not differ between the crosslinked and uncrosslinked receptors (Furuichi et

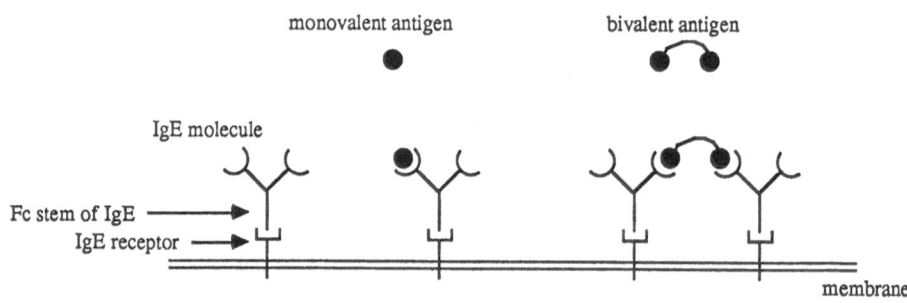

FIGURE 6.1 Crosslinking of IgE receptors. When monovalent IgE receptors are bound by the Fc stem of IgE molecules, the receptors become functionally bivalent due to the presence of two identical binding sites on the IgE molecule. The IgE receptors can then be crosslinked via these IgE molecules if a bivalent antigen binds to two IgE molecules simultaneously.

al. 1985), the effect of the ligand is apparently on the sorting outcome.

Clearly, the sorting process can provide a wide range of outcomes. Yet we believe that a single fundamental physical/chemical mechanism may underlie all of these possibilities. In an Chapter 4, we examined the sorting process theoretically and found that the general features of sorting can be accounted for by a mechanism based on ligand and receptor diffusion within endosomes along with trapping of receptors. This separation scheme, while able to explain the "typical" receptor-ligand sorting scheme in which most receptors are recycled and most ligands are degraded, does not predict the different outcomes of the sorting process found upon changes in the receptor and ligand properties.

We now extend the earlier theory to account for the wide range of possible outcomes observed for the sorting process. The sorting mechanism we propose predicts that the outcome of the sorting process is a function of the ability of the ligand molecules to alter the rate of transport of receptors from the central vesicle of the endosome into the tubules.

6.2 Mathematical Model

We make several assumptions in this model for receptor/ligand sorting in the endosome. The beginning of the sorting process is defined as the time that internalized receptor-ligand complexes are first located in the endosome, a low pH environment with the geometry of a central vesicle and attached thin tubules. We assume that the vesicle and tubule dimensions do not change during sorting. We also assume that all receptors and ligands are initially located in the vesicle, presumably the result of the fusion of several smaller vesicles formed at the cell surface (Braell 1987; Murphy 1985). The sorting process ends with the detachment of the tubule(s) from the central vesicle at a time we define as the sorting time. Any receptors and ligands in tubules at the end of this sorting time are assumed to be recycled, and molecules remaining behind in the vesicle are assumed to be eventually degraded (Geuze et al. 1983; Hatae et al. 1986; Geuze et al. 1987).

Previous investigators and the theory of Chapter 4 have assumed that during sorting at the low endosome pH, most receptors and their ligands completely dissociate (Gex-Fabry and DeLisi 1984). However, experimental measurements indicate that this may not always be the case. Measurements of the pH-dependent dissociation of receptor-ligand complexes indicate that the dissociation rate constant for the complex at pH 5.0 to pH 5.5 may not be great enough to release all ligands in the endosome from their receptors within the sorting time (Mellman et al. 1984; Dunn and Hubbard 1984; Wileman et al. 1985b). In addition, ligands which dissociate may rebind to another receptor, as recently demonstrated by Schiff et al. (1984). Thus many of the ligands may remain bound to receptors within the endosome, at least for some substantial fraction of the total sorting time. We feel that the inclusion of this fact is a key element in explaining the different effects of ligands on the sorting process. Thus the model includes the dissociation and rebinding of ligands and their receptors; multivalent ligands can also crosslink or aggregate receptors within the endosome. These events are shown schematically in Figure 6.2 for two cases, monovalent receptors with either monovalent or bivalent ligands. The rate constants k_1 and k_{-1}, as in the whole cell model, denote the binding and dissociation of ligands and their receptors. However, these rate constants must now be evaluated at the low endosome pH. The rate constants for crosslinking and release of crosslinked receptors at the low pH are given by k_2 and k_{-2}, respectively. The values of the per receptor rate constants k_1, k_{-1}, k_2, and k_{-2} can be estimated from data on the binding of ligands to whole cells; one method is given in Appendix II. Conditions in the vesicle and tubule(s) will be assumed identical and therefore the rate constants for binding and dissociation in the two compartments are the same. We will also assume that the endosome pH is approximately constant throughout the sorting process so that these rate constants are not time-dependent.

The movement of receptors and ligands within the endosome can also occur. In particular, we are interested in the movement of molecules from the vesicle into one of the tubules because these are the molecules that are recycled. We assume that receptors can diffuse within the plane of the endosome membrane, as they do on the cell surface (Schlessinger et al. 1978; Wiegel 1984). In the previous modeling (Chapter 4), we found that the highly efficient receptor recycling, or sorting of receptors into tubules, found in some systems (Geuze et al. 1983; Dunn and Hubbard 1984; Marshall 1985b) can be explained by postulating the existence of a trapping mechanism in the tubule. The trapping mechanism may be the interaction of receptors in tubules with other molecules, such as clathrin, a similar trapping molecule, or elements of the cytoskeleton, serving to trap and effectively concentrate receptors prior to recycling. In our model, we will assume that receptors diffusing into tubules are trapped there and cannot diffuse back into the vesicle. In Chapter 4, we calculated the rate constant for this movement, both with and without a membrane convective current.

We assume that ligands not bound to receptors are free to diffuse within and between the tubule and vesicle volumes. The rate constant for the movement of ligands into tubules has been calculated and shown to be 10 to 100 times greater than that for receptor movement into tubules by pure diffusion (see Chapter 4). The rate constant for the movement of ligand molecules out of the tubule should also be also great. Thus in the time required for an appreciable number of receptors to accumulate in a tubule, the ligand molecules will have diffused into and out of the tubule many times. For this reason, we will assume that the distribution of free ligand molecules between the vesicle and tubule volumes of the endosome is always at equilibrium. Knowledge of this equilibrium distribution must include the calculation of the partition coefficient, κ, to correct for excluded volume in the tubule due to the

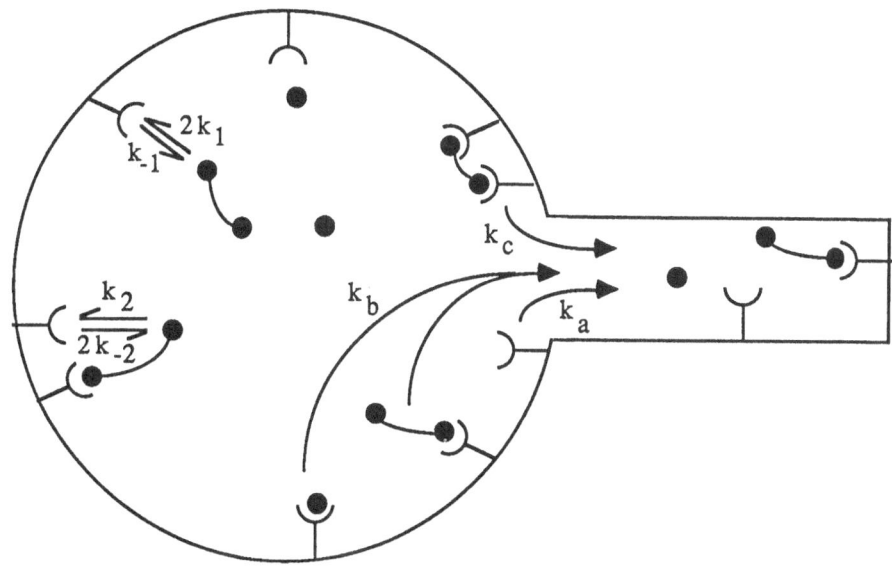

FIGURE 6.2 Single endosome model. For simplicity, only one tubule is shown connected to the central vesicle of the endosome. Receptors and ligands can exist in bound or free states; bivalent ligands can crosslink two receptors. The relevant rate constants k_1, k_{-1}, k_2, and k_{-2} must be evaluated at the low pH of the endosome. Receptors may move by passive diffusion or with the aid of a membrane current into the tubule, where they are trapped and cannot return to the vesicle. Ligand molecules may diffuse between the vesicle and tubule compartments. Other parameters are defined in the text.

finite size of the ligand molecule (Chapter 4 and Pappenheimer 1953).

As detailed in the introduction, experiments with several receptor systems show that a change in ligand properties may alter the outcome of the sorting process. Clearly, changes in ligand affinity and valency will affect the fractions of receptors that are free, bound, or crosslinked in the endosome at all times. We therefore suggest that the ability of a receptor to move from the vesicle into a tubule is a function of the receptor state: free, bound (meaning bound by ligand but not crosslinked), or crosslinked. This is shown in Figure 6.2 by the assignment of three different rate constants for receptor movement, k_a for free receptors, k_b for bound receptors, and k_c for crosslinked receptors. What we propose, as suggested by the experimental evidence, is that crosslinked or crosslinked and bound receptors are hindered in their entry into the tubule, or that k_c alone or both k_c and k_b may be significantly less than k_a. Note that because receptors are assumed to be trapped in tubules, there are no rate constants assigned to the movement of receptors from the tubule back into the vesicle.

We can suggest two possible explanations for a decreased ability of bound and/or crosslinked receptors to move into tubules as compared with free receptors. First, there may be an interaction between complexes and other molecules, such as elements of the cytoskeleton, which hinders the movement of the complexes and therefore their ability to diffuse into the tubules. Menon et al. (1986a, 1986b) have used fluorescence photobleaching techniques to measure the mobilities of bound and crosslinked IgE receptors on the cell surface of rat basophilic leukemia (RBL) cells. They found that receptors bound by monomeric IgE molecules moved with a diffusion coefficient of 3×10^{-10} cm^2/sec., but that the diffusion coefficient of receptors crosslinked with IgE trimers and higher oligomers is reduced to a value less than or equal to 5×10^{-12} cm^2/sec. Because Robertson and co-workers (Robertson et al. 1986) have shown that cell surface receptors become detergent insoluble along with the cytoskeleton only after crosslinking, this change in receptor mobility is likely the result of interactions between the crosslinked complexes and the cytoskeleton. Most recently, this group has found that clustered receptors become associated with a constitutively-insoluble component on the RBL cell surface (B. Baird, personal communication).

The probable association of bound surface receptors with cytoskeletal elements has been described in two other systems, EGF receptors on A431 cells (Wiegant et al. 1986) and α-interferon receptors on human Daudi lymphoblastoid cells (Pfeffer et al. 1987). Finally, crosslinked class I major hisocompatibility proteins on Epstein-Barr virus-transformed B cells have been shown to diffuse more slowly than when not crosslinked, possibly due to cytoskeletal interactions (Bierer et al. 1987). Interestingly, this effect was not seen in a second cell type, interferon-γ-treated human dermal fibroblasts, suggesting that different cell types may have different rates of or different types of cytoskeletal interactions for the same receptor-ligand system. We suggest that similar interactions may be occurring in the endosome and may be responsible for slowing the movement of bound and/or crosslinked receptors.

The degree to which receptors in different states might be slowed in their diffusion would depend upon several factors. The affinity of the receptors in their bound and free states for an interaction molecule may be different if, for example, a conformational change in the receptor upon binding increases its affinity for this interaction molecule. There is, in fact, recent evidence for some type of conformational change in the mannose 6-phosphate receptors upon ligand binding (Westcott et al. 1987). Secondly, if the affinity of bound receptors for the interaction molecule is low, an increase in the number of receptors in an aggregate will increase the probability that at least one receptor is bound to the interaction molecule, thus immobilizing the entire aggregate. The same is true if the interaction molecule is multivalent. In addition, in this case a low affinity receptor/interaction molecule binding could be overcome by the presence of multiple attachment sites on a receptor aggregate. Some of these possibilities are shown schematically in Figure 6.3.

In these scenarios, then, the transport rate constants for the movement of receptors in each of the various bound states would be functions of the association and dissociation rate constants for binding of receptors to the interaction molecule, the number of binding sites on the interaction molecule and on a receptor aggregate, and the concentrations of receptors and interaction molecules. All of these factors may differ among various cell types and among different receptor systems in the same cell type, thus providing a possible explanation of the basis of different sorting outcomes.

As a second possibility for the decreased ability of receptors in some states to move into tubules, we suggest that some complexes may be too large to enter the thin tubules. For example, a complex containing two receptors, each projecting about 75 Å from the membrane, and a ligand molecule, perhaps about 30 Å in diameter, may be unable to enter a tubule as narrow as 100 Å in diameter although a single unbound receptor would not be as hindered. Receptor, ligand, and tubule dimensions may vary among different systems and would, as in the first scenario, provide a basis for understanding the different sorting outcomes.

Using the rate constants detailed above for the binding, dissociation, and transport events in the endosome, we can describe the evolution with time of the populations of free, bound, and crosslinked receptors and ligands in the vesicle and tubule(s) of the endosome. The partitioning of receptors and ligands between the vesicle and tubule at the end of the sorting time gives the separation obtained in the sorting process and therefore the fractions of endocytosed receptors and ligands routed to the lysosomes, f_R and f_L.

We consider two cases: monovalent receptors bound by monovalent ligands and monovalent receptors bound by bivalent ligands.

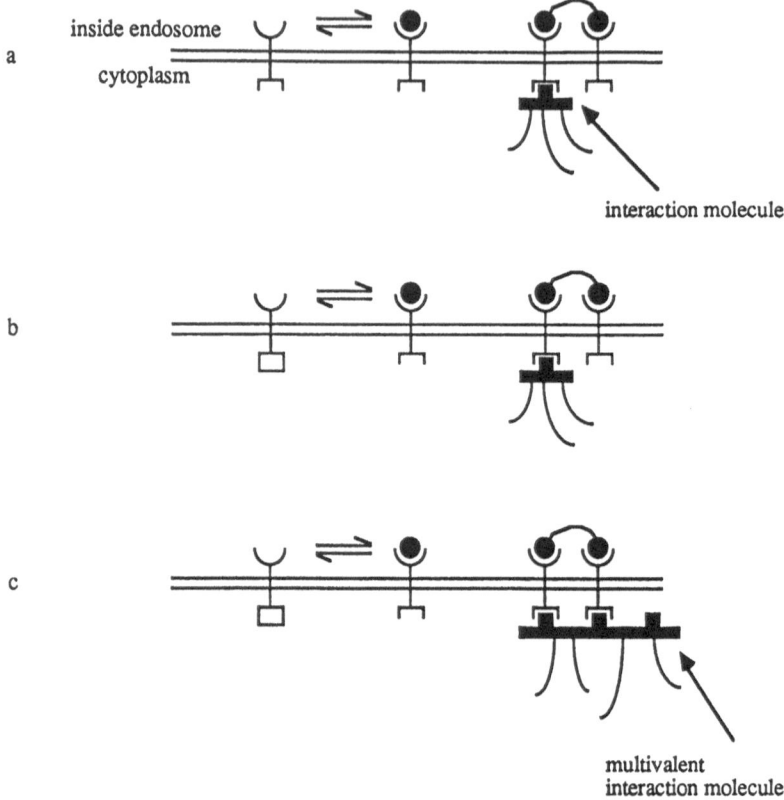

inside endosome

cytoplasm

a

interaction molecule

b

c

multivalent
interaction molecule

FIGURE 6.3 Possible scenarios which may determine the values of $\gamma 1$ and $\gamma 2$. Changes in the bound state of a receptor may result in a change in the rate of transport of that receptor into tubules. (a) Receptors have a low affinity for a monovalent interaction molecule. Ligand binding produces no conformational change in the receptor, but a multivalent binding of ligand to receptors increases the probability that at least one of the receptors in the aggregate, and therefore the entire aggregate, will be immobilized. Thus $\gamma 1 \approx 1$ and $\gamma 2 < \gamma 1$. (b) Free receptors have no affinity for a monovalent interaction molecule. Ligand binding produces a conformational change in the receptor so that it may bind the interaction molecule. If the affinity of the ligand-bound receptor for the interaction molecule is great enough, $\gamma 1 < 1$. $\gamma 2$ may be significantly less than $\gamma 1$ because the probability that one of two crosslinked receptors is bound to the interaction molecule and thus both receptors are immobilized is even greater than the probability that any single ligand-bound receptor is immobilized. (c) Same as case b except that the interaction molecule is multivalent. If the affinity of a ligand-bound receptor for the interaction molecule is low, $\gamma 1 \approx 1$. Significantly, $\gamma 2$ wil be less than $\gamma 1$ because of the enhanced affinity of the multivalent interaction molecule for the aggregate due to the presence of a greater number of binding sites. If the affinity of a ligand-bound receptor for the interaction molecule is high, then both $\gamma 1$ and $\gamma 2$ are much less than 1.

Case I—Monovalent receptor, Monovalent ligand

Three linearly independent differential equations which describe changes in the number of receptor and ligand molecules in each state (free, bound, or crosslinked) and each location (vesicle or tubule) with time are:

$$\frac{d R_f}{d t} = -k_1 L_f R_f + k_{-1} L_1 - k_a R_f \tag{6.1}$$

$$\frac{d L_1}{d t} = k_1 L_f R_f - k_{-1} L_1 - k_b L_1 \tag{6.2}$$

$$\frac{d L_1'}{d t} = k_1 L_f' R_f' - k_{-1} L_1' + k_b L_1, \tag{6.3}$$

subject to the conservation equations for the total number of receptor and ligand molecules:

$$R_T = R_f + R_f' + R_1 + R_1' \tag{6.4}$$

$$L_T = N_A (V_V L_f + V_T L_f') + L_1 + L_1' \tag{6.5}$$

and the equation giving the instantaneous equilibrium of free ligand molecules between the vesicle and tubule(s):

$$L_f' = \kappa L_f \tag{6.6}$$

R_f, R_1, and L_1 are the number of free receptors, bound receptors, and bound ligands, respectively, in the vesicle (number/endosome). L_f is the concentration of free ligand in the vesicle (moles/liter). R_f', R_1', L_1', and L_f' represent the same quantities as the unprimed variables, but for the tubules. Because both the receptors and ligands are monovalent, R_1 can be replaced by L_1 and R_1' by L_1'. R_T is the total number of receptors in the endosome, and L_T is the total number of ligand molecules in the endosome. V_V and V_T are the vesicle and tubule volumes; N_A is Avogadro's number. The partition coefficient κ is given by

$$\kappa = (1 - \lambda)^2, \tag{6.7}$$

where λ is equal to the ratio of the diameter of the ligand molecule to the diameter of the tubule (Pappenheimer 1953).

We define the dimensionless variables $\overline{R}_f$, $\overline{R}_f'$, $\overline{L}_1$, and $\overline{L}_1'$ as the original variables R_f, R_f', L_1, and L_1' divided by the total number of receptors R_T. The dimensionless time τ is defined by

$$\tau = k_a t . \tag{6.8}$$

Upon algebraic manipulation, three equations in the dimensionless variables result:

$$\frac{d\overline{R}_f}{d\tau} = -\varepsilon [\eta \overline{R}_f (\rho - \overline{L}_1 - \overline{L}_1') - \overline{L}_1] - \overline{R}_f \tag{6.9}$$

$$\frac{d\overline{L}_1}{d\tau} = \varepsilon [\eta \overline{R}_f (\rho - \overline{L}_1 - \overline{L}_1') - \overline{L}_1] - \gamma 1 \, \overline{L}_1 \tag{6.10}$$

$$\frac{d\overline{L}_1'}{d\tau} = \varepsilon [\eta \kappa (\rho - \overline{L}_1 - \overline{L}_1')(1 - \overline{R}_f - \overline{L}_1 - \overline{L}_1') - \overline{L}_1'] + \gamma 1 \, \overline{L}_1 \tag{6.11}$$

with the dimensionless groups defined by:

$$\varepsilon = \frac{k_{-1}}{k_a} \tag{6.12}$$

$$\eta = (\frac{k_1}{k_{-1}}) \frac{R_T}{N_A (V_V + \kappa V_T)} \tag{6.13}$$

$$\rho = \frac{L_T}{R_T} \tag{6.14}$$

$$\gamma 1 = \frac{k_b}{k_a} . \tag{6.15}$$

The parameter ε is a comparison of dissociation and transport kinetics, η is the dimensionless equilibrium constant, ρ compares the number of ligand and receptors molecules present in a single endosome, and $\gamma 1$ is a transport parameter relating the rates of movement of free and bound receptors.

As stated earlier, we assume that all receptors are initially located in the vesicle. We must also specify the fraction of internalized receptors which are bound as sorting begins, f_b. Thus in dimensionless form the initial conditions are

$$\bar{\bar{R}}_f(0) = 1 - f_b \tag{6.16}$$

$$\bar{\bar{L}}_1(0) = f_b \tag{6.17}$$

$$\bar{\bar{L}}_1'(0) = 0. \tag{6.18}$$

We make the simplifying assumption that receptor/ligand complexes are not significantly effected by any gradual lowering of pH on their way to the endosome and that therefore the initial condition in the endosome can be predicted from information about the complexes at the moment of internalization. If only bound receptors are internalized and delivered to the endosome, f_b is equal to one. If instead one assumes that the binding of extracellular ligand to cell surface receptors reaches equilibrium and that receptors are endocytosed and delivered to the endosome in states proportional to their existence on the cell surface, then f_b is equal to the fraction of receptors which would be bound at equilibrium and is a function of only the extracellular ligand concentration L_0 and the equilibrium constant for binding at the cell surface K_1^N. In this second case, then,

$$f_b = \frac{K_1^N L_0}{1 + K_1^N L_0}, \tag{6.19}$$

where K_1^N is equal to k_1^N / k_{-1}^N.

The total amount of ligand present in the endosome, L_T, is important in determining the value of the dimensionless group ρ. L_T includes not only ligand molecules entering the endosome bound to receptors but also those internalized nonspecifically in the bulk fluid. If we assume that the volume of fluid internalized at the cell surface is equal to the volume of the vesicle and that free ligand was internalized in that volume at a concentration equal to the extracellular concentration L_0, then the value of ρ is found from

$$\rho = \bar{\bar{L}}_1(0) + \frac{N_A V_v L_0}{R_T}. \tag{6.20}$$

Eqs. 6.9 - 6.11 can be solved with the initial conditions given in Eqs. 6.16-6.18 to obtain $\overline{R_f}$, $\overline{L_1}$, and $\overline{L_1}'$. The sorting fractions f_R and f_L can then be found from

$$f_R = \overline{\overline{R}}_f + \overline{\overline{L}}_1 \tag{6.21}$$

$$f_L = \frac{\left(\dfrac{V_V}{V_V + \kappa V_T}\right)(\rho - \overline{\overline{L}}_1 - \overline{\overline{L}}_1') + \overline{\overline{L}}_1}{\rho} \tag{6.22}$$

Case II—Monovalent receptor, Bivalent ligand

When a ligand molecule is bivalent, it can bind to two receptors at once, crosslinking them. Because We propose that sorting depends on the distinction between bound and crosslinked receptors, each receptor state must be followed separately.

Five linearly independent differential equations to describe the endosome populations for this case are:

$$\frac{dR_f}{dt} = -2k_1 L_f R_f + k_{-1} L_1 - k_2 L_1 R_f + 2k_{-2} L_2 - k_a R_f \tag{6.23}$$

$$\frac{dL_1}{dt} = 2k_1 L_f R_f - k_{-1} L_1 - k_2 L_1 R_f + 2k_{-2} L_2 - k_b L_1 \tag{6.24}$$

$$\frac{dL_1'}{dt} = 2k_1 L_f' R_f' - k_{-1} L_1' - k_2 L_1' R_f' + 2k_{-2} L_2' + k_b L_1 \tag{6.25}$$

$$\frac{dL_2}{dt} = k_2 L_1 R_f - 2k_{-2} L_2 - k_c L_2 \tag{6.26}$$

$$\frac{dL_2'}{dt} = k_2 L_1' R_f' - 2k_{-2} L_2' + k_c L_2' . \tag{6.27}$$

Note that a factor of two is included in the initial binding of solution ligand to a receptor and in the dissociation of a crosslinked ligand from one of the two receptors it binds. This is a statistical factor necessary because in each of these cases there are two identical events which can occur due to the bivalent nature of the ligand. These equations must be solved along with the conservation equations on the total number of receptor and ligand molecules in the endosome

$$R_T = R_f + R_f' + L_1 + L_1' + 2L_2 + 2L_2' \tag{6.28}$$

$$L_T = N_A(V_V L_f + V_T L_f') + L_1 + L_1' + L_2 + L_2' \tag{6.29}$$

and Eq. 6.6 from Case I. L_1 and L_1' are the numbers of ligand molecules bound to only one receptor in the vesicle and tubule(s), respectively. L_2 and L_2' are the numbers of ligand molecules crosslinking two receptors in the vesicle and tubule(s). All other variables are as defined previously. Note that the total number of bound receptors in the endosome is the sum of the bound receptors in the vesicle, $L_1 + 2L_2$, and the bound receptors in the tubule, $L_1' + 2L_2'$.

The new variables L_2 and L_2' are made dimensionless by

$$\bar{\bar{L}}_2 = \frac{2L_2}{R_T} \tag{6.30}$$

$$\bar{\bar{L}}_2' = \frac{2L_2'}{R_T}. \tag{6.31}$$

The remaining variables are scaled as in Case I. With these scalings and the appropriate algebraic manipulations, the following equations in the dimensionless variables result:

$$\frac{d\bar{\bar{R}}_f}{d\tau} = \epsilon[-2\eta \bar{\bar{R}}_f(\rho - \bar{\bar{L}}_1 - \bar{\bar{L}}_1' - 0.5\bar{\bar{L}}_2 - 0.5\bar{\bar{L}}_2') + \bar{\bar{L}}_1$$

$$- \chi \bar{\bar{R}}_f \bar{\bar{L}}_1 + \sigma \bar{\bar{L}}_2] - \bar{\bar{R}}_f \tag{6.32}$$

$$\frac{d\bar{\bar{L}}_1}{d\tau} = \epsilon[2\eta \bar{\bar{R}}_f(\rho - \bar{\bar{L}}_1 - \bar{\bar{L}}_1' - 0.5\bar{\bar{L}}_2 - 0.5\bar{\bar{L}}_2') - \bar{\bar{L}}_1$$

$$- \chi \bar{\bar{R}}_f \bar{\bar{L}}_1 + \sigma \bar{\bar{L}}_2] - \gamma 1 \bar{\bar{L}}_1 \tag{6.33}$$

$$\frac{d\bar{\bar{L}}_1'}{d\tau} = \epsilon[\,2\eta\kappa(\rho - \bar{\bar{L}}_1 - \bar{\bar{L}}_1' - 0.5\bar{\bar{L}}_2 - 0.5\bar{\bar{L}}_2')(1 - \bar{\bar{R}}_f - \bar{\bar{L}}_1$$

$$-\bar{\bar{L}}_1' - \bar{\bar{L}}_2 - \bar{\bar{L}}_2') - \bar{\bar{L}}_1' - \chi\bar{\bar{L}}_1'(1 - \bar{\bar{R}}_f - \bar{\bar{L}}_1 - \bar{\bar{L}}_1' - \bar{\bar{L}}_2 - \bar{\bar{L}}_2')$$

$$+\sigma\bar{\bar{L}}_2'\,] + \gamma 1\,\bar{\bar{L}}_1 \tag{6.34}$$

$$\frac{d\bar{\bar{L}}_2}{d\tau} = \epsilon[\,2\chi\bar{\bar{R}}_f\bar{\bar{L}}_1 - 2\sigma\bar{\bar{L}}_2\,] - \gamma 2\,\bar{\bar{L}}_2 \tag{6.35}$$

$$\frac{d\bar{\bar{L}}_2'}{d\tau} = \epsilon[\,2\chi\bar{\bar{L}}_1'(1 - \bar{\bar{R}}_f - \bar{\bar{L}}_1 - \bar{\bar{L}}_1' - \bar{\bar{L}}_2 - \bar{\bar{L}}_2') - 2\sigma\bar{\bar{L}}_2'\,] + \gamma 2\,\bar{\bar{L}}_2\,, \tag{6.36}$$

with the additional dimensionless groups defined by:

$$\chi = \frac{k_2}{k_{-1}}R_T \tag{6.37}$$

$$\sigma = \frac{k_{-2}}{k_{-1}} \tag{6.38}$$

$$\gamma 2 = \frac{k_c}{k_a}\,. \tag{6.39}$$

χ represents the dimensionless rate of crosslinking and σ the rate of breakup of those crosslinks. The transport parameter $\gamma 2$ compares the rates of movement of crosslinked and free receptors.

We will again assume that all receptors and ligands are initially located in the vesicle. The fraction of receptors which are initially bound but uncrosslinked, f_b, and crosslinked, f_c, must be specified. As in the monovalent ligand case, it is simplest to assume that the initial states of receptors are related to their states on the cell surface and that binding at the cell surface reaches equilibrium prior to endocytosis. The theory of Perelson and DeLisi (1980) predicts that when the binding of a bivalent ligand to monovalent surface receptors reaches equilibrium, the fractions of bound but uncrosslinked surface receptors, f_b, and crosslinked receptors, f_c, are found from:

$$f_b = \beta \left[\frac{-1 + \sqrt{1 + 4\delta}}{2\delta} \right] \qquad (6.40)$$

$$f_c = \left[\frac{1 + 2\delta - \sqrt{1 + 4\delta}}{2\delta} \right], \qquad (6.41)$$

where the dimensionless parameters β and δ are given by

$$\beta = \frac{2 K_1^N L_0}{1 + 2 K_1^N L_0} \qquad (6.42)$$

$$\delta = \beta (1 - \beta) K_2^N S_0. \qquad (6.43)$$

K_2^N is the equilibrium constant for crosslinking, k_2^{N*}/k_{-2}^N, and S_0 is the density of surface receptors (receptors/area). It is most important to note here the difference between k_2^* (with superscript N if evaluated for normal or extracellular pH) and k_2, the crosslinking rate constant used inside the endosome. The rate constant k_2^* has units of area/time and is used when receptor densities are given in receptors/area, as in Eq. 6.43. The rate constant k_2, however, has units of time^{-1} and is used for the equations governing behavior inside the endosome because we have chosen to express the receptor concentrations as receptors/endosome. In general, the conversion from one set of units to the other is made by considering the relevant surface area; in our case, the relevant surface area is that of the endosome, A_{end}, so that $k_2 = k_2^*/A_{end}$.

The equilibrium crosslinking equations, Eqs. 6.40 and 6.41, give the result that the number of crosslinked receptors on the cell surface is small at low ligand concentrations, increases to a maximum as the ligand concentration increases, and then decreases as the ligand concentration is increased still further. Because of this maximum in the equilibrium crosslinking curve, found at $2K_1^N L_0$ equal to one, increasing the concentration of bivalent ligand in solution may increase or decrease the number of crosslinked receptors internalized, depending on the ligand concentration, equilibrium constants, and surface receptor density. A sample equilibrium crosslinking curve is given in Figure 6.4.

To examine the sorting outcome when receptors are internalized and delivered to the

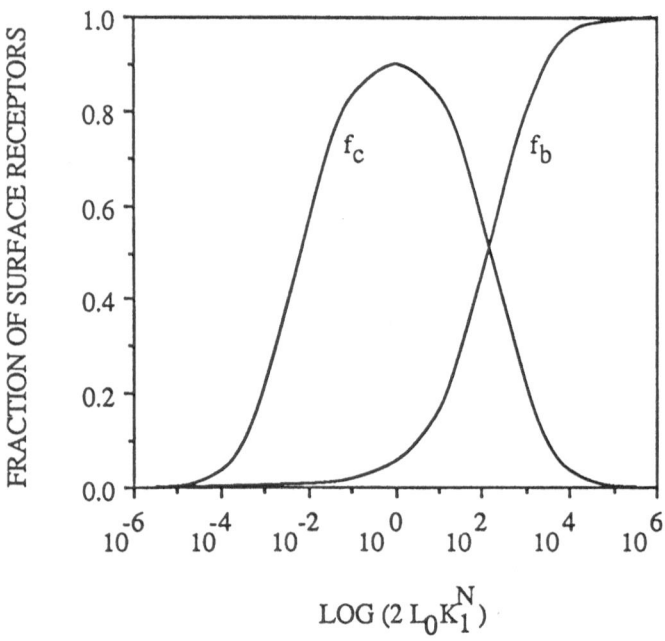

FIGURE 6.4 Equilibrium crosslinking curve. The fractions of crosslinked receptors, f_c, and bound (but not crosslinked) receptors, f_b, are shown. Parameter values: $K_1N = 1 \times 10^6$ M, $K_2N = 1 \times 10^{-8}$ cm^2, $S_0 = 3.18 \times 10^{10}$ receptors/cm^2.

endosome in proportion to their state on the cell surface, the initial conditions in dimensionless form are

$$\overline{\overline{R}}_f(0) = 1 - f_b - f_c \tag{6.44}$$

$$\overline{\overline{L}}_1(0) = f_b \tag{6.45}$$

$$\overline{\overline{L}}_1'(0) = 0 \tag{6.46}$$

$$\overline{\overline{L}}_2(0) = f_c \tag{6.47}$$

$$\overline{\overline{L}}_2'(0) = 0. \tag{6.48}$$

If instead no free receptors are internalized, then $\overline{R}_f(0)$ is equal to zero, $\overline{L_1}(0)$ is equal to the ratio $f_b/(f_b+f_c)$ and $\overline{L_2}(0)$ is equal to the ratio $f_c/(f_b+f_c)$. As an example, if $L_0 = 10^{-8}$ M, $K_1N = 10^8$ M^{-1}, $K_2N = 2.0 \times 10^{-6}$ cm^2, and $S_0 = 3.18 \times 10^{10}$ receptors/cm^2 (10^5 receptors on a spherical cell with radius 5 μm) and no free receptors are endocytosed, then $\overline{L_1}(0) = 0.01$ and $\overline{L_2}(0) = 0.99$.

As in Case I, the value of ρ and the initial states of the receptors are not independent. If we again assume that the volume of the vesicle is equal to the volume of fluid internalized and that this fluid contained ligand at the same concentration as in the medium, the value of ρ can be found from

$$\rho = \overline{\overline{L}}_1(0) + 0.5\,\overline{\overline{L}}_2(0) + \frac{L_0 N_A V_V}{R_T}. \tag{6.49}$$

Eqs. 6.32 - 6.36 can be solved to obtain $\overline{R}_f$, $\overline{L}_1$, and $\overline{L}_1'$, $\overline{L}_2$, and $\overline{L}_2'$. The sorting fractions f_R and f_L can then be found from

$$f_R = \overline{\overline{R}}_f + \overline{\overline{L}}_1 + \overline{\overline{L}}_2 \tag{6.50}$$

$$f_L = \frac{(\dfrac{V_V}{V_V + \kappa V_T})(\rho - \overline{\overline{L}}_1 - \overline{\overline{L}}_1' - 0.5\overline{\overline{L}}_2 - 0.5\,\overline{\overline{L}}_2') + \overline{\overline{L}}_1 + 0.5\,\overline{\overline{L}}_2}{\rho}. \tag{6.51}$$

The model equations describing the sorting process in Case I and Case II were solved numerically (Gear 1971). The predictions of the model were examined for a variety of parameter values.

6.3 Model Predictions

Reasonable parameter ranges

In order to examine the predictions of the model, it is first helpful to specify physiologically reasonable ranges for several of the model parameters. The rate constant for the movement of free receptors from the vesicle to a tubule when there is no membrane current was calculated in Chapter 4 as

$$k_a = \frac{D_R}{R^2} \left[\frac{2 \ln \left[\frac{2}{1 - \cos \theta_c} \right]}{1 + \cos \theta_c} - 1 \right]^{-1},$$ (6.52)

where D_R is the diffusion coefficient of the receptor in the endosome membrane, R is the radius of the vesicle, b is the radius of the tubule, and θ_c is the ratio b/R. For the reasonable values $D_R = 1 \times 10^{-10}$ cm^2/sec, R = 0.3 µm, and b = 0.02 µm (Geuze et al. 1983; Marsh et al. 1986; Wiegel 1984), the rate constant k_a is on the order of 1 min^{-1}. In examining the predictions of the model, therefore, it is reasonable to assume that for many systems the dimensionless time τ is approximately equal to the sorting time in minutes. Overall, given the realistic ranges for R, 1×10^{-5} to 4×10^{-5} cm., for b, 5×10^{-7} to 3×10^{-6} cm., and a fixed value for D_R, 1×10^{-10}, the rate constant k_a will range from 0.41 min^{-1} to 21. min^{-1}. The value of the diffusion coefficient D_R has been fixed, instead of varied over the range specified in Chapter 5, to reflect the narrow range of diffusion coefficients that have been measured in the types of receptor systems to which the model can be applied (Menon et al. 1986; Schlessinger et al. 1978; Hillman and Schlessinger, 1982; Giugni et al. 1987). If the sorting time is on the order of 5 to 10 min., then the dimensionless sorting time may extend up to a value of 210.

The association and dissociation rate constants for monovalent receptor and monovalent ligand binding are available for several systems, typically at normal pH but occasionally also at the low endosome pH. Some of these values are given in Table 6.1. Thus a reasonable range of k_1 for study may be 1×10^6 M^{-1}min^{-1} to 1×10^8 M^{-1}min^{-1} and k_{-1} may vary in the range of 1×10^{-3} min^{-1} to 1 min^{-1}. As examined in Appendix 2, k_2 and k_{-2} are difficult to measure directly. Dembo et al. (1982) used a model to fit association, dissociation, and crosslinking

number	receptor	ligand	cell type	S_0	pH 7.2		
					K_1 (M^{-1})	k_1 (M^{-1}min^{-1})	k_{-1} (min^{-1})
1	Fcγ receptor	Fab fragment of anti-receptor antibody 2.4G2	mouse macrophage	7.1×10^5	1.3×10^9	3×10^6	2.3×10^{-3}
2	chemotactic peptide receptor	FNLLP[c]	rabbit peritoneal polymorphonuclear leukocyte	5×10^4	5×10^7	2×10^7	0.4
3	mannose receptor	mannose/BSA	rabbit alveolar macrophage	—	—	—	—
4	asialoglyco-protein receptor	ASOR[d]	HepG2[e]	1.5×10^5 - 2.5×10^5	—	2.23×10^6	< 0.01
5	transferrin receptor	transferrin[f]	K562[g]	1.5×10^5	5.3×10^8	—	—
6	transferrin receptor	transferrin[f]	HepG2[e]	5×10^4	2.85×10^7 - 3.36×10^7	3.02×10^6	0.09 - 0.106
7	insulin receptor	insulin	liver plasma membrane	—	5×10^8 - 1×10^{10}	1.2×10^8	0.012 - 0.24
8	insulin receptor	insulin	rat adipocyte	3×10^5	—	—	—
9	epidermal growth factor receptor	epidermal growth factor	perfused rat liver	1.5×10^5	5×10^8 - 1×10^9	—	—

TABLE 6.1 Literature data on association and dissociation constants for several receptor/ligand systems. Data is given at both normal and low pH, when available. [a] Except as noted. [b] Calculated from graphical data assuming first order dissociation kinetics. [c] N-formlynorleucylleucylphenylalanine. [d] Asialoorosomucoid (a multivalent ligand). [e] A human hepatoma cell line. [f] Values given at pH 7.2 are for diferric transferrin; values given at pH 4.8 are for apotransferrin. [g] A human leukemic cell line. Table is continued on the following page.

pH 5.0[a]

number	K_1 (M^{-1})	k_1 ($M^{-1}min^{-1}$)	k_{-1} (min^{-1})	references
1	—	—	0.1[b]	Mellman and Unkeless 1980 Mellman et al. 1984
2	5×10^7	2×10^7	0.4	Zigmond et al. 1982
3	—	—	0.46[b]	Wileman et al. 1985b
4	—	—	—	Ciechanover et al. 1985
5	4.8×10^7 (pH 4.8)	2×10^6 (pH 4.8)	0.041 (pH 4.8)	Klausner et al. 1983a
6	—	—	—	Ciechanover et al. 1983
7	—	—	—	Corin and Donner 1982
8	—	—	—	Marshall 1985a
9	—	—	0.46[b]	Dunn and Hubbard 1984

data and obtained a value of $k_2{}^*$ for the IgE receptor (Fc_ε) system on human basophils of approximately 5×10^{-10} cm^2/sec at pH 7.2. Using the value of endosome area of 1.5×10^{-8} cm^2, as found by Marsh et al. (1986) for BHK-21 cells, one can obtain a value for k_2 of 2.0 min^{-1}. Although this measurement was not taken at the low endosome pH, it is often assumed that dissociation and not association kinetics are most affected by a change in pH, and therefore we suspect that k_2 in the endosome is quite close to this value. The value of k_2 in many other systems may be within one or two orders of magnitude of this value. Lastly, we anticipate that k_{-2} will be of the same order of magnitude as k_{-1}. One might also estimate the relevant ranges of k_2 and k_{-2} from the ranges for k_1 and k_{-1} just given by using the methods in Appendix 2 and reasonable values for the receptor radius, number of receptors in the endosome, and receptor and ligand diffusion coefficients.

The total number of receptors in the endosome, R_T, has not to our knowledge been measured. The density of receptors on the cell surface can be estimated from the total surface receptor number, 5×10^4 to 1×10^6 receptors/cell, and a typical cell diameter, 1×10^{-3} cm., to be 1.6×10^{10} to 3.2×10^{11} receptors/cm^2. If the density of receptors in the endosome is in the range of one-half to ten times this density and the endosome area is 1.5×10^{-8} cm^2 (Marsh et al. 1986), then R_T ranges from 120 to 4800 receptors/endosome. Marsh et al. (1986) also determined that approximately 60 to 70% of the total volume of the endosome, i.e., 60 to 70% of 4×10^{-14} cm^3 in BHK-21 cells, is contained in the vesicle, giving values for V_V and V_T.

The value for the partition coefficient κ can be estimated from a typical ligand diameter, 10 to 50 Å, and tubule diameter, 100 to 600 Å, to be in the range of 0.25 to 0.97.

Given the above estimates of several of the fundamental parameters, we can calculate estimated physiologic ranges for the model parameters. In doing so, it can be noted that the dimensionless group χ differs from η only in the substitution of the rate constant k_2 for $k_1/(N_A(V_V + \kappa V_T))$. One can calculate by the methods of Appendix 2 that k_2 varies over approximately the same range as $k_1/(N_A(V_V + \kappa V_T))$. In addition, the ranges are "matched" in that a low value of the association rate constant k_1 will dictate a corresponding low value of the crosslinking rate constant k_2 because several of the same biochemical/biophysical interactions occur in both reactions. Thus we anticipate that the two dimensionless groups η and χ will be roughly of the same order of magnitude. The model parameters, definitions, physical meanings, and ranges of values are listed in Table 6.2.

Several of the model parameters can be seen to vary over a broad region. This reflects in part an uncertainty in the value of several fundamental parameters, such as k_a and R_T. The ranges are also broad, however, because of the variety of reaction kinetics that have been shown to exist in physiological systems, as shown in Table 6.1.

model parameter	mathematical definition	physical meaning	estimated physiologic range
τ	$k_a t$	dimensionless time	$0 - 210^a$
ε	$\dfrac{k_{-1}}{k_a}$	ratio of dissociation rate constant at endosome pH to free receptor transport rate constant	$4.8 \times 10^{-5} - 2.4$ [a]
ρ	$\dfrac{L_T}{R_T}$	ratio of total ligand molecules in endosome to total receptor molecules in endosome	depends on choice of initial condition
κ	$(1-\lambda)^2$	partition coefficient (measure of the effect of ligand size on its equilibrium concentration in tubules)	0.25 - 0.97
η	$\left(\dfrac{k_1}{k_{-1}}\right)\dfrac{R_T}{N_A(V_V + \kappa V_T)}$	dimensionless equilibrium constant at endosome pH	$5.0 - 2.9 \times 10^7$
χ	$\dfrac{k_2}{k_{-1}} R_T$	dimensionless rate of crosslinking at endosome pH	roughly the same order of magnitude as η
σ	$\dfrac{k_{-2}}{k_{-1}}$	dimensionless rate of breakup of crosslinks at endosome pH	0.1 - 10
$\gamma 1$	$\dfrac{k_b}{k_a}$	ratio of transport rate constant for singly bound receptors to that of free receptors	0 - 1
$\gamma 2$	$\dfrac{k_c}{k_a}$	ratio of transport rate constant for crosslinked receptors to that of free receptors	0 - 1
f_b	—	for monovalent ligand case only, the fraction of receptors in the endosome that are bound when sorting begins	0 - 1
$\ddot{\bar{R}}_f(0)$	see Eqs. 6.40, 6.41, and 6.44 - 48	for bivalent ligand case only, the fraction of receptors in the endosome that are unbound when sorting begins	0 - 1
$\ddot{\bar{L}}_1(0)$		for bivalent ligand case only, the fraction of receptors in the endosome that are singly bound when sorting begins	0 - 1
$\ddot{\bar{L}}_2(0)$		for bivalent ligand case only, the fraction of receptors in the endosome that are crosslinked when sorting begins	0 - 1

TABLE 6.2 Model parameters, their definitions, and their expected range of values. The estimation of the parameter ranges is given in the text. The last four parameters in the table are used to specify the initial conditions for the model. Note that $\bar{R}_f(0) + \bar{L}_1(0) + \bar{L}_2(0) = 1$. [a] No membrane current.

General behavior of the system

Using our model, it is possible to follow the evolution of the different receptor populations in the endosome and to monitor the progress of sorting as a function of time. Sample results are shown in Figure 6.5a for the case of monovalent receptors bound by monovalent ligands and Figure 6.5b for the case of monovalent receptors bound by bivalent ligands.

The initial conditions for the monovalent receptor, monovalent ligand separation shown in Figure 6.5a reflect the situation in which only bound receptors are internalized and delivered to the endosome. The parameter $\gamma 1$ has been set to a value of 0.1, and therefore bound receptors are hindered in their movement into tubules. Receptors, initially bound and located in the vesicle, slowly move into and accumulate within tubules. There is a transient increase in free receptors in the vesicle (curve d), facilitated by the higher value of the dissociation rate constant at low pH found in many systems. The appearance of bound receptors in tubules (curve b) reflects binding events that have occurred in the tubules as well as the movement of bound receptors directly into tubules. When the sorting process ends, these receptors in tubules will presumably be sent along the recycling pathway in this bound state. Whether or not these receptors remain bound when they reach the cell surface will depend on the conditions of their intracellular transport: the pH of the recycling compartment, the concentrations of receptors and ligands in this compartment, and the time required to reach the cell surface.

A similar plot is shown in Figure 6.5b for the case of monovalent receptors bound by bivalent ligands. Here, the initial conditions have been chosen so that 75% of the receptors enter the sorting chamber crosslinked and the remaining 25% enter bound. The transport parameters $\gamma 1$ and $\gamma 2$ are in this case set equal to 1 and 0, respectively, so that singly bound receptors can enter the tubules as rapidly as free receptors but crosslinked receptors cannot enter at all.

In both cases, all receptors eventually accumulate in the tubules due to the trapping mechanism. The kinetics of this accumulation vary with the values of the dimensionless parameters ε, η, ρ, $\gamma 1$, χ, σ, $\gamma 2$, and τ and the initial conditions in the endosome. To predict the actual outcome of the sorting step, the value of the dimensionless sorting time τ^*, the total time allowed by the cell for the sorting process, is needed. As stated earlier, a sorting time of 5 to 10 minutes is consistent with much of the experimental data.

These types of plots, detailing the populations of receptors in free, bound, or crosslinked states and in each location, can be made for any desired combination of parameter values. In order to efficiently examine the effects of many of the dimensionless parameters and to relate these results to the outcome of the sorting process, all future results will present only the

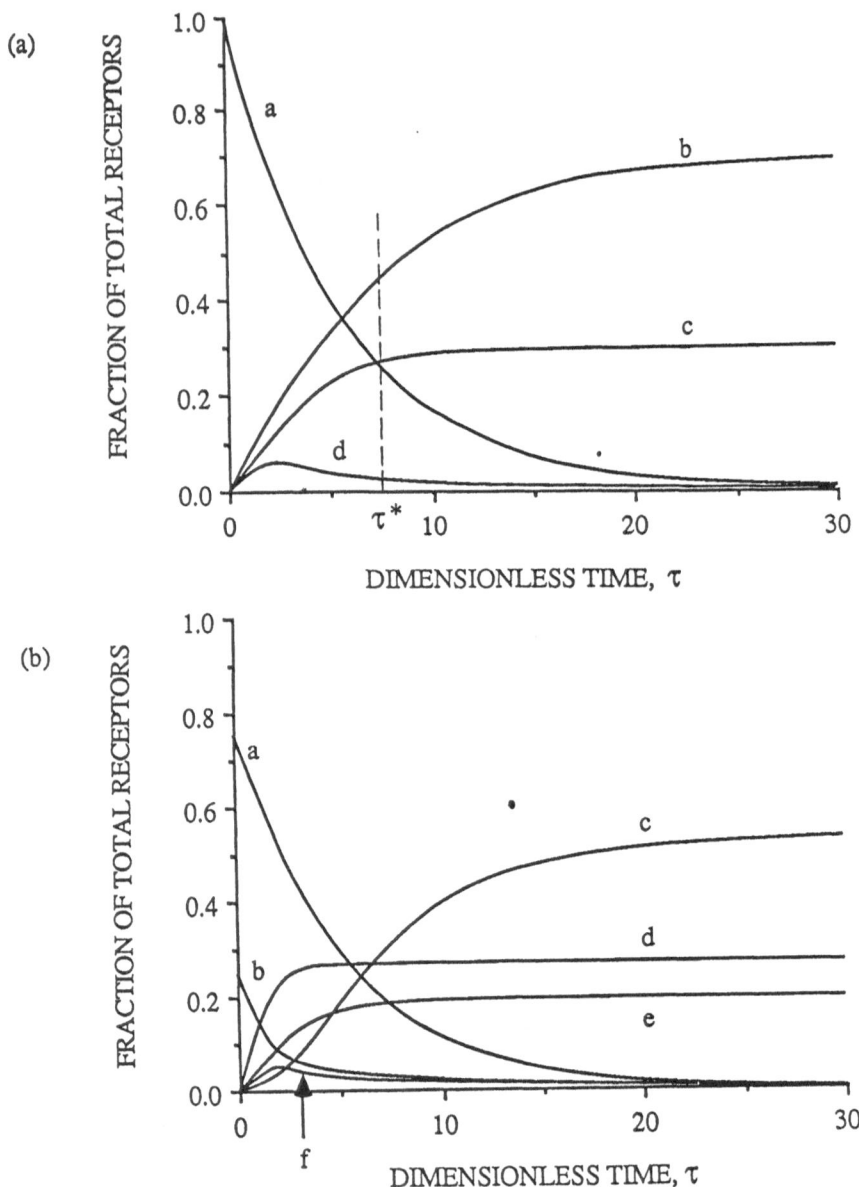

FIGURE 6.5 General behavior of the single endosome model. (a) Monovalent receptor, monovalent ligand system. Parameter values: $\varepsilon = 0.1$, $\eta = 10.0$, $\rho = 1.001$, $\kappa = 0.75$, $\gamma 1 = 0.1$, $f_b = 1.0$. Curve a: bound receptors in vesicle; curve b: bound receptors in tubules; curve c: free receptors in tubules; curve d: free receptors in vesicle. (b) Monovalent receptor, bivalent ligand system. Parameter values: $\varepsilon = 0.1$, $\eta = 10.0$, $\chi = 10.0$, $\sigma = 1.0$, $\rho = 0.63$, $\kappa = 0.75$, $\gamma 1 = 1.0$, $\gamma 2 = 0.0$, $\overline{R}_f (0) = 0.0$, $\overline{L}_1 (0) = 0.25$, $\overline{L}_2 (0) = 0.75$. Curve a: crosslinked receptors in vesicle; curve b: bound receptors in vesicle; curve c: crosslinked receptors in tubule; curve d: bound receptors in tubule; curve e: free receptors in tubule; curve f: free receptors in vesicle.

variation of the sorting fractions, f_R and f_L. These lumped quantities can be obtained from the individual population results using Eqs. 6.21 and 6.22 for Case I and Eqs. 6.50 and 6.51 for Case II.

Effect of transport rates

We have postulated that the rate at which receptors enter tubules may depend on the receptor state: free, bound, or crosslinked. This dependence may be due to an interaction of bound and/or crosslinked receptors with a component of the cytoskeleton or to steric effects, among other possibilities. Our suggestions mean that the transport parameters $\gamma 1$ and $\gamma 2$ may take on a value less than one. The actual values of these two parameters will depend on factors not known at this time, such as the affinity of free and bound receptors for cytoskeletal interaction proteins and the concentrations of these proteins, all of which may vary between different cell types and different receptor systems in the same cell. In Fig. 6.6a, the effect of changing the value of $\gamma 1$ when the ligand is monovalent is demonstrated for fixed values of the other parameters. The variation in the sorting fraction f_R is shown as a function of time. A similar plot is shown in Figure 6.6b for the case of bivalent ligands. Here, $\gamma 1$ is set to its maximum value of one so that bound receptors are not hindered in their entry into a tubule, and $\gamma 2$ is allowed to vary. The results show that the kinetics of receptor accumulation in tubules are drastically slowed when the transport parameters are reduced from their maximum value of one. Given a constant value of the sorting time, then, the overall rate constant for receptor recycling, $k_R (1-f_R)$, will be lowered as the transport parameter(s) are reduced.

Effect of extracellular ligand concentration

A change in the extracellular ligand concentration L_0 can affect the fraction of receptors recycled, $1- f_R$, in two ways for the monovalent receptor, monovalent ligand system. The amount of ligand internalized nonspecifically in the bulk fluid of a vesicle pinching off at the cell surface and delivered to the endosome along with receptors will change with L_0, thus changing the ratio of ligand to receptor molecules, ρ (Eq. 6.20). The effects of a change in ρ when all receptors are initially bound ($f_b = 1.0$) are seen in Figure 6.7a. Note that substantial changes in the ligand concentration are required to produce a significant effect on receptor recycling. For example, if $R_T = 1000$ receptors and $V_V = 2.8 \times 10^{-17}$ liters, a change in ligand concentration from 10^{-8} to 10^{-5} M would be required to change the value of ρ from 1.0002 to 1.2. If the other parameters are fixed as stated in the legend of Figure 6.7a and the sorting time τ^* is equal to 6, this increase in ρ would reduce the fraction of receptors recycled only from 0.24 to 0.16. This small change, however, is significant when the resulting effect

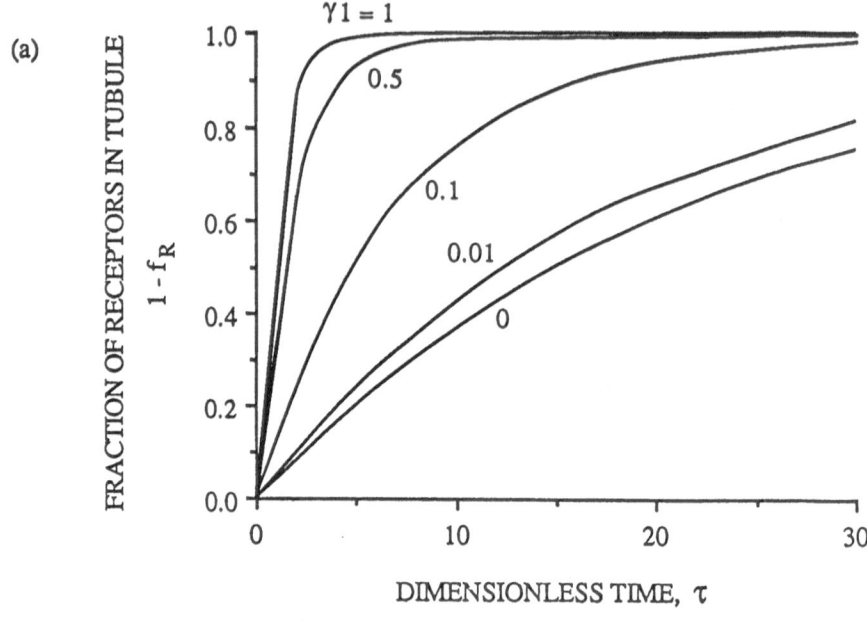

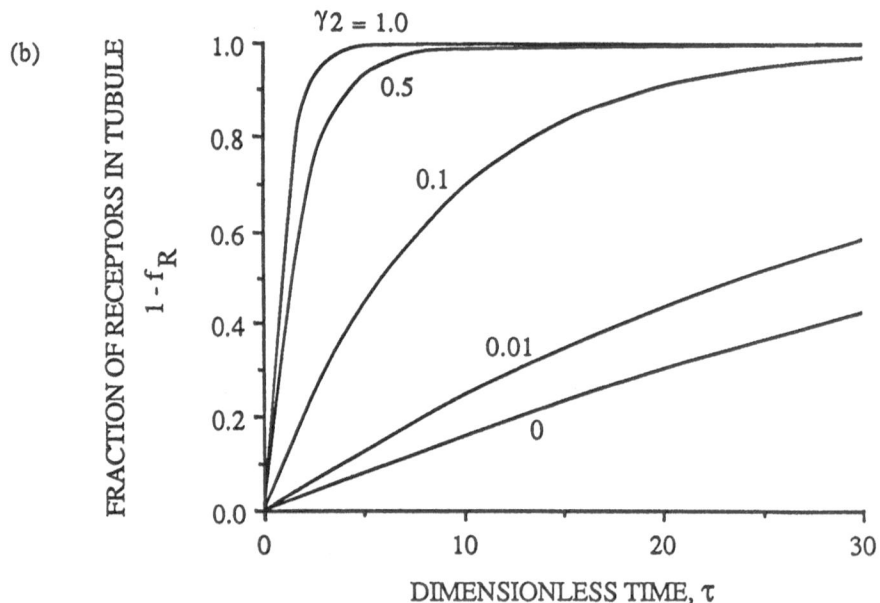

FIGURE 6.6 Effect of transport parameters on receptor recycling. For both plots, $\varepsilon = 0.1$, $\eta = 100.0$, $\kappa = 0.75$. (a) Monovalent receptor, monovalent ligand system. Parameter values: $\rho = 1.001$, $f_b = 1.0$. (b) Monovalent receptor, bivalent ligand system. Parameter values: $\chi = 100.0$, $\sigma = 0.1$, $\rho = 0.51$, $\gamma 1 = 1.0$, $\overline{R}_f(0) = 0.0$, $\overline{L}_1(0) = 0.01$, $\overline{L}_2(0) = 0.99$.

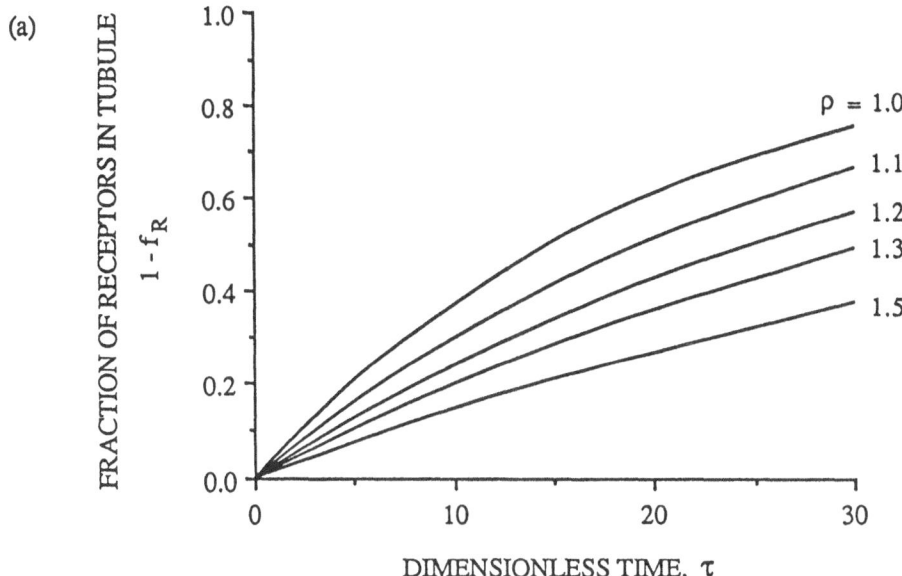

FIGURE 6.7 Effect of extracellular ligand concentration L_0 on receptor recycling. For all four plots, $\varepsilon = 0.1$, $\eta = 100.0$, $\kappa = 0.75$. (a) Monovalent receptor, monovalent ligand system. In this plot only ρ is affected by a change in the extracellular ligand concentration. Parameter values: $\gamma 1 = 0$, $f_b = 1.0$. (b) Monovalent receptor, monovalent ligand system. Here a change in L_0 affects both ρ and f_b. For $K_1^N = 5 \times 10^7$ M^{-1}, $V_V = 2.8 \times 10^{-17}$ liters, and $R_T = 1000$ receptors/endosome, the extracellular ligand concentrations and their corresponding f_b and ρ values are: for $L_0 = 10^{-9}$ M, $f_b = 0.048$, and $\rho = 0.048$; for $L_0 = 10^{-8}$ M, $f_b = 0.333$, and $\rho = 0.334$; for $L_0 = 10^{-7}$ M, $f_b = 0.833$, and $\rho = 0.835$; for $L_0 = 10^{-6}$ M, $f_b = 0.980$, and $\rho = 0.997$. For all curves, $\gamma 1 = 0$. (c) Monovalent receptor, bivalent ligand system. Although no free receptors are internalized, a change in L_0 affects both ρ and the initial distribution of bound and crosslinked receptors. For $K_1^N = 1 \times 10^6$ M^{-1}, $K_2^N = 1 \times 10^{-8}$ cm^2, $V_V = 2.8 \times 10^{-17}$ liters, $R_T = 1000$ receptors/endosome, and $S_0 = 3.18 \times 10^{10}$ receptors/cm^2, extracellular ligand concentrations and their corresponding $\overline{L}_1(0)$, $\overline{L}_2(0)$, and ρ values are: for $L_0 = 10^{-9}$ M, $\overline{L}_1(0) = 0.005$, $\overline{L}_2(0) = 0.995$, and $\rho = 0.502$; for $L_0 = 10^{-6}$ M, $\overline{L}_1(0) = 0.078$, $\overline{L}_2(0) = 0.922$, and $\rho = 0.556$; for $L_0 = 10^{-5}$ M, $\overline{L}_1(0) = 0.222$, $\overline{L}_2(0) = 0.778$, and $\rho = 0.780$. For all curves, $\chi = 100.0$, $\sigma = 1.0$, $\gamma 1 = 1.0$, $\gamma 2 = 0.0$, and $\overline{R}_f(0) = 0.0$. (d) Monovalent receptor, bivalent ligand system. For this plot, free, bound, and crosslinked receptors are internalized, and a change in L_0 affects ρ and the three initial conditions. For the same values of K_1^N, K_2^N, R_T, and S_0 as in part c, the extracellular ligand concentrations and the corresponding ρ, $\overline{R}_f(0)$, $\overline{L}_1(0)$, and $\overline{L}_2(0)$ values are: for $L_0 = 10^{-9}$ M, $\overline{L}_1(0) = 0.001$, $\overline{L}_2(0) = 0.306$, and $\rho = 0.154$; for $L_0 = 10^{-6}$ M, $\overline{L}_1(0) = 0.075$, $\overline{L}_2(0) = 0.888$, and $\rho = 0.536$; for $L_0 = 10^{-5}$ M, $\overline{L}_1(0) = 0.220$, $\overline{L}_2(0) = 0.769$, and $\rho = 0.773$. Remaining parameter values are as in part c, except that $\overline{R}_f(0) = 1.0 - \overline{L}_1(0) - \overline{L}_2(0)$.

(b)

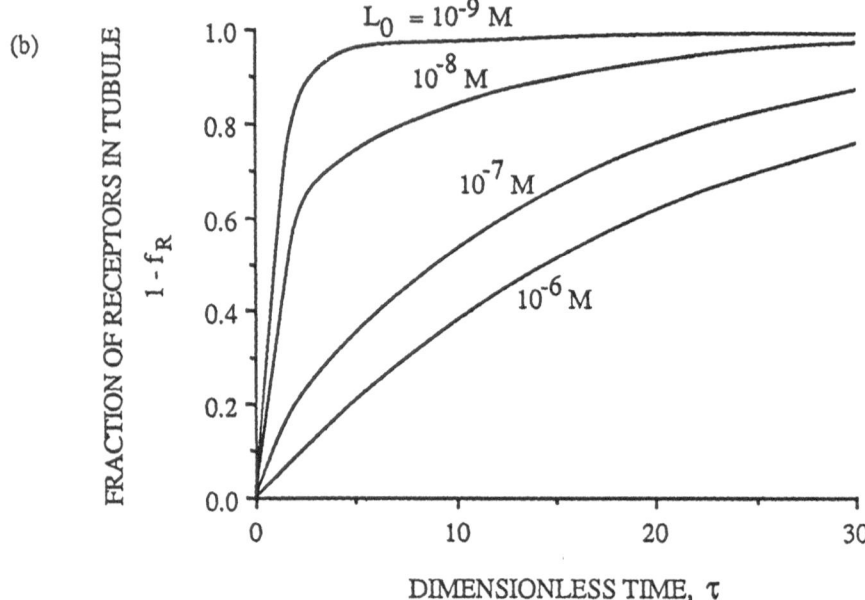

(c)

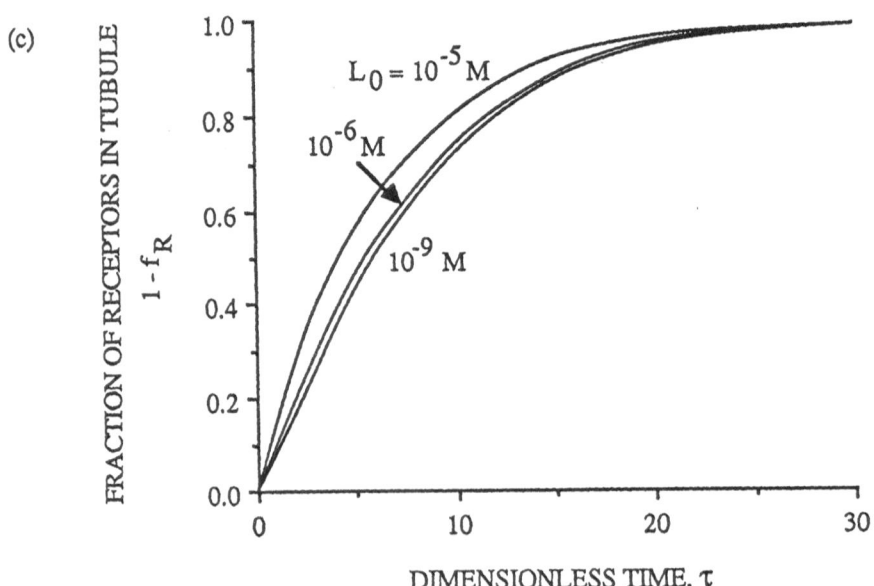

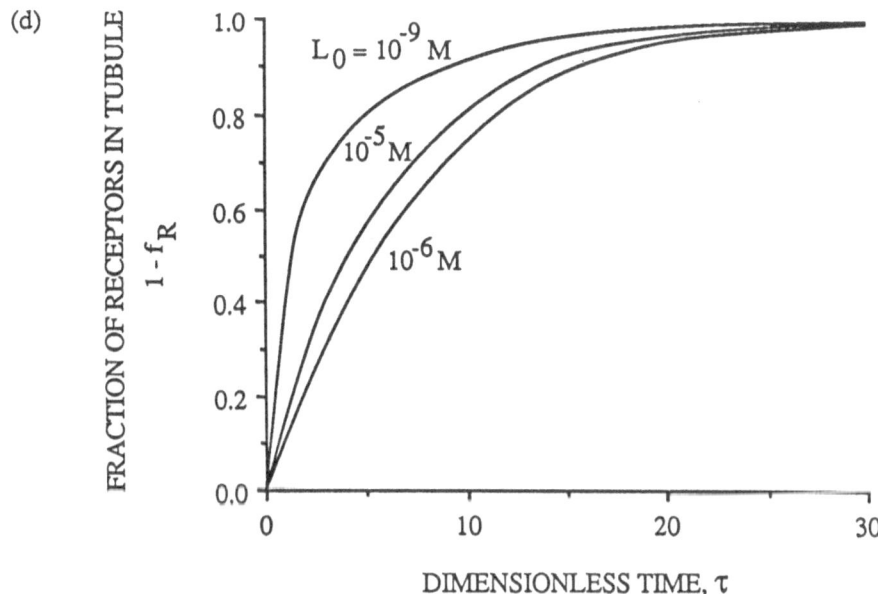

on total receptor number over many cycles of endocytosis is considered.

Secondly, if both bound and free receptors are internalized, changes in the extracellular ligand concentration can affect the predicted separation of receptors and ligands by changing the initial conditions through the parameter f_b as well as by changing the value of ρ. If we assume that the fraction of receptors bound initially in the endosome is the same as the fraction that would be bound at the cell surface at equilibrium, the effects of a change in ligand concentration L_0 on sorting can be calculated. A sample plot is shown in Figure 6.7b, where the fraction of receptors bound at time zero was calculated from Eq. 6.19 and the parameter ρ was calculated from Eq. 6.20. In contrast to Figure 6.7a, in which only the value of ρ varied with L_0, changes in extracellular ligand concentration of only one order of magnitude can now have a significant effect on the sorting outcome.

For the monovalent receptor, bivalent ligand system, the effects of a change in the extracellular ligand concentration L_0 can be examined in a similar fashion. In Figure 6.7c, we plot the effect of a change in L_0 over five orders of magnitude when only bound and crosslinked receptors are internalized. The initial conditions are calculated from $\overline{L}_1(0) = f_b/(f_b + f_c)$ and $\overline{L}_2(0) = f_c/(f_b + f_c)$ where f_b and f_c are functions of L_0. Because the initial conditions do not vary greatly even over this large change in L_0, the separation in the endosome is not altered significantly. Note, however, that an increase in L_0 increases the fraction of receptors recycled. This occurs because the calculated proportion of receptors entering the endosome crosslinked decreases as the ligand concentration is increased for the parameter values chosen for this figure.

In Figure 6.7d, we consider the same monovalent receptor, bivalent ligand system but allow free receptors to enter the endosome also, in proportion to their existence on the cell surface at equilibrium. The initial conditions for this situation are given by Eqs. 6.44-6.48. A change in the extracellular ligand concentration now has a greater effect, because for the same change in L_0 as in Figure 6.7c, the changes in ρ and the initial conditions are over a wider range of values when all three populations of receptors, free, bound, and crosslinked, are endocytosed. An interesting feature of this plot is that the curve for $L_0 = 10^{-5}$ M actually lies between the curves for two lower ligand concentrations. The reason for this is clear if one recalls the biphasic nature of the equilibrium crosslinking curve described by Eqs. 6.40 and 6.41. For the parameter values given in the figure legend and for $L_0 = 10^{-9}$ M, the number of crosslinked receptors falls on the rising portion of the equilibrium crosslinking curve (see Figure 6.4). When the ligand concentration is increased to 10^{-6} M, the maximum in the crosslinking curve, which occurs at 5×10^{-7} M, has been passed and the number of crosslinked receptors lies on the falling portion of the curve. However, the number of crosslinked receptors is greater than at the much lower ligand concentration of 10^{-9} M. As the ligand concentration is increased once more, to 10^{-5} M, the number of crosslinked receptors is less than at $L_0 = 10^{-6}$ M but greater than at $L_0 = 10^{-9}$ M. This increase and then decrease in the number of crosslinked receptors initially present in the endosome as the ligand concentration is increased is responsible for the decrease and then increase in the kinetics of receptor accumulation within the tubules.

Affinity effects

A key assumption of our model is that receptors in different bound states move into tubules at different rates. Because of this assumption, the rate constants for ligand binding, dissociation, and crosslinking play a major role in governing the separation of receptors and ligands. In Figure 6.8 we show plots which give more complete information on the separation by indicating both the fractions of receptors and ligand which are recycled as a function of the dimensionless sorting time. For the case of monovalent ligands with ε equal to 0.1 and γ1 equal to 0.1, Figure 6.8a shows the effect of the equilibrium parameter η on both the fractions of receptor and ligand molecules recycled. Note that both the time course and the value of η for each separation are shown. Clearly, changes in the affinity of a ligand for a receptor are predicted to have a significant effect on both the amount of ligand and receptor recycled: both the efficiency and the kinetics of the separation vary with η. The efficiency of the separation is determined by locating the point on the plot corresponding to the particular values of the sorting time and affinity for the receptor-ligand system in question and reading from the axes the fractions of ligand and receptors recycled. Points falling in the lower right hand region of the plot indicate a fairly "efficient" separation; the fraction of receptors recycled is high and the fraction of ligand recycled is low. On the other hand, points falling on the diagonal from $1-f_L = 1-f_R = 0$ to $1-f_R = 1-f_L = 1$ indicate a "poor" separation. In this case, so many of the receptors in the tubule are bound that there is essentially no separation of the two populations of molecules. The kinetics of the separation are indicated by following the progression of a separation along a curve of constant η. One can see that the accumulation of receptors in tubules proceeds more rapidly for η = 10, for example, than for η = 100. It is interesting to note that the curve for η = 0.1, unlike the curves for greater affinity, has a maximum. This occurs when the kinetics of bound receptor accumulation in tubules are more rapid than the kinetics of ligand dissociation from these receptors. Bound receptor molecules may move into a tubule, albeit at a slower rate then free receptors. At early times, the tubules contain both this bound ligand and a fixed fraction of the free ligand (Eq. 6.6). As the time increases, the amount of ligand in the tubule begins to decrease because most of the bound ligand dissociates (recall that the affinity is very low) and the tubule cannot hold more than this same fixed fraction of the free ligand.

A similar plot is shown for the monovalent ligand case in Figure 6.8b, but the value of ε

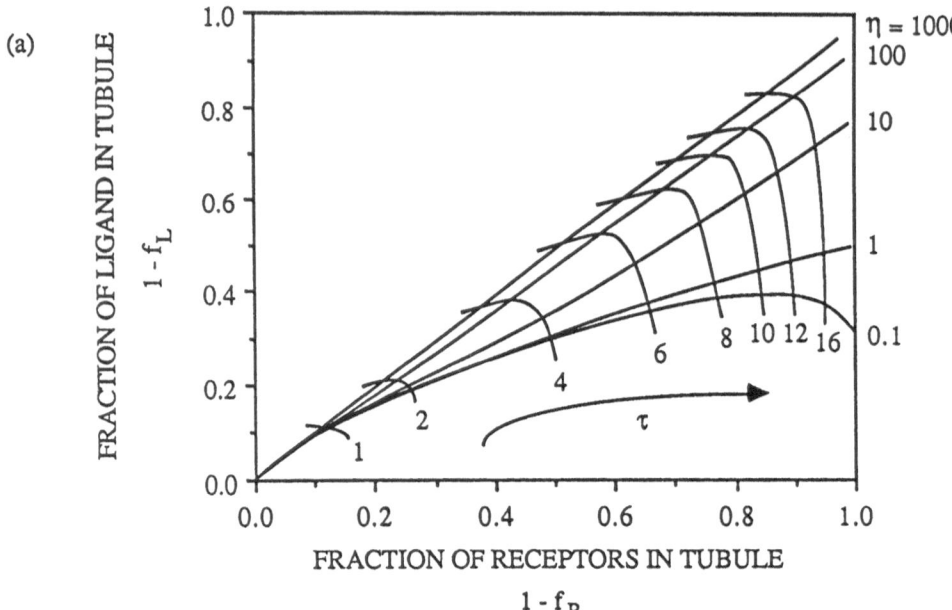

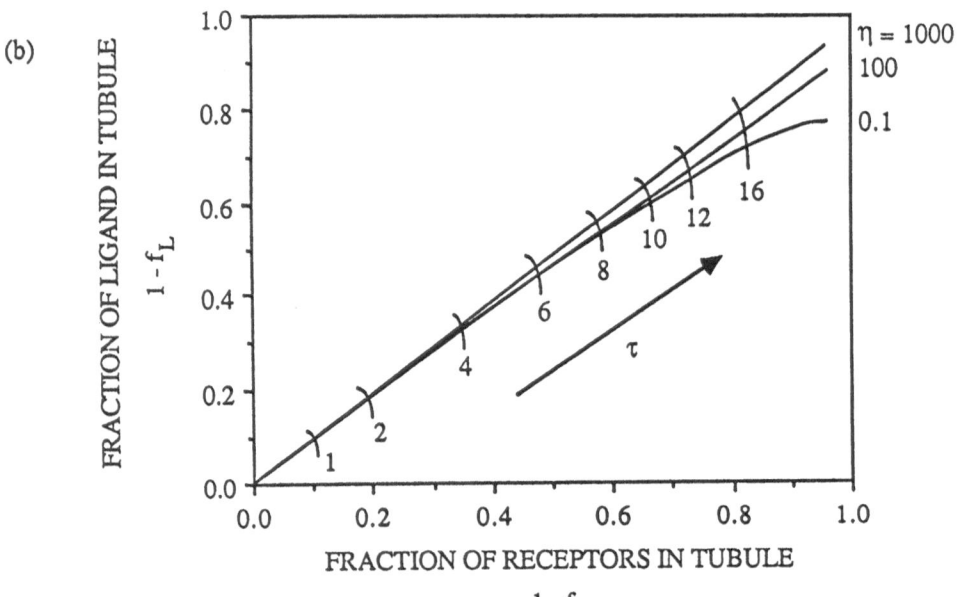

FIGURE 6.8 Effect of affinity on receptor and ligand recycling. For all four plots, κ = 0.75, V_V/V_T = 7/3. (a) Monovalent receptor, monovalent ligand system. Parameter values: ε = 0.1, $\gamma 1$ = 0.1, ρ = 1.001, f_b = 1.0. (b) Monovalent receptor, monovalent ligand system. Parameter values are as in a except that ε = 0.01. (c) Monovalent receptor, bivalent ligand system. Parameter values: ε = 1.0, σ = 1.0, ρ = 0.63, $\gamma 1$ = 1.0, $\gamma 2$ = 0.0, $\overline{R}_f(0)$ = 0.0, $\overline{L}_1(0)$ = 0.25, $\overline{L}_2(0)$ = 0.75. (d) Monovalent receptor, bivalent ligand system. Parameter values are as in c except that ε = 0.01.

(c)

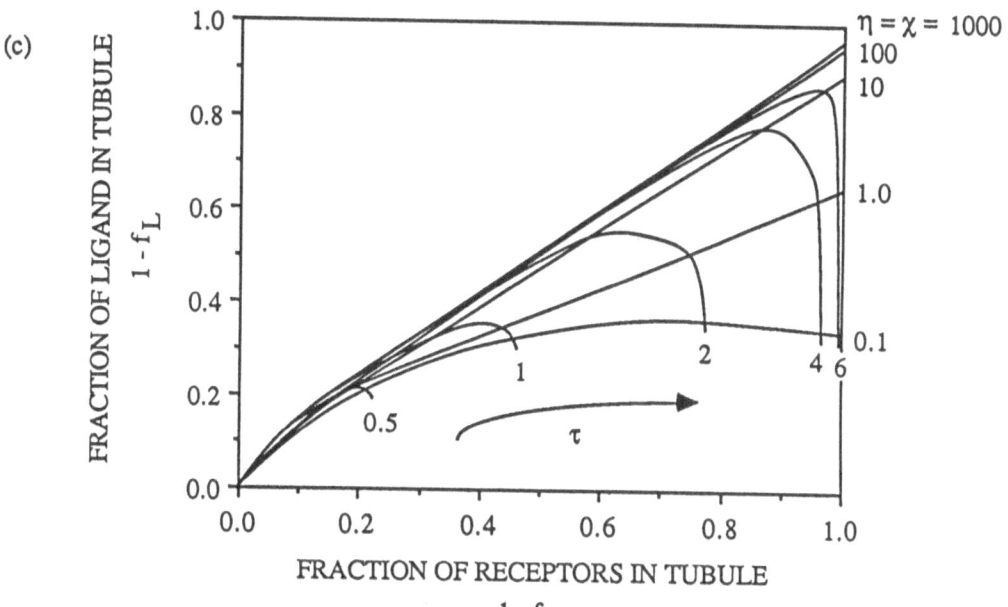

(d)

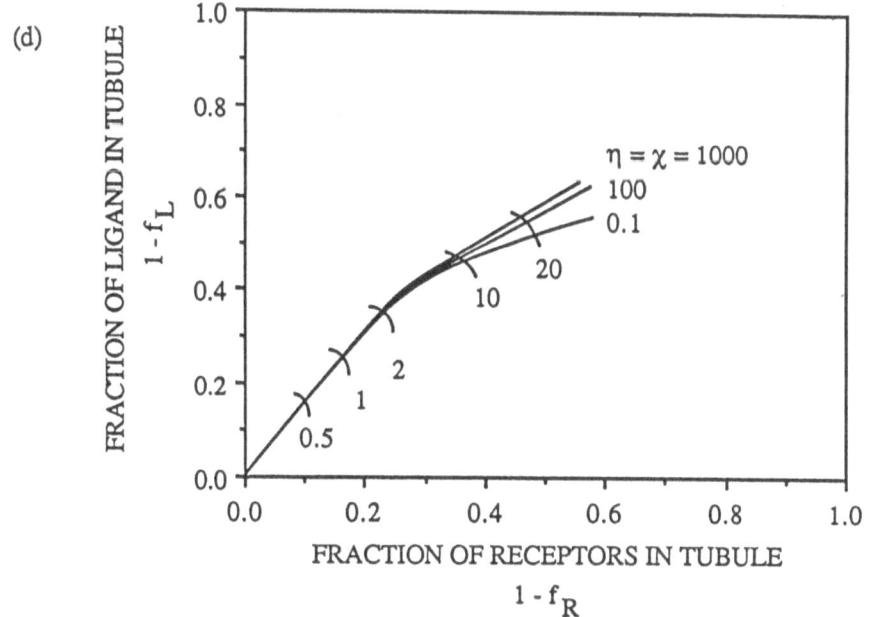

has been reduced to 0.01. In contrast to Figure 6.8a, a change in the equilibrium parameter η has essentially no effect on the outcome of the separation. Also note that the time courses of each separation are slowed over those shown in the previous figure. The reason for this is clear: a reduction in the value of ε may mean that the ligands are much less pH sensitive (k_{-1} is reduced) and therefore little dissociation of receptor and ligand is expected in a few minutes. A second interpretation is that the transport rate constant k_a is increased, indicating that for the same dimensionless time as in Figure 6.8a fewer minutes have elapsed. If this is the case, for a dimensionless time of one in Figure 6.8a the same real time is denoted by a dimensionless time of 10 in Figure 6.8b.

Similar plots are shown in Figures 6.8c and 6.8d for the monovalent receptor, bivalent ligand case. For the parameter values given in the legend of the figure and for $\varepsilon = 0.1$, the influence of affinity (through the parameters η and χ) on the separation is great. However, for $\varepsilon = 0.01$ there is very little effect of affinity.

Most importantly, Figure 6.8 shows that it is possible to see different effects of affinity in different systems. Depending on the relevant parameter ranges, the endosome may be operating in a regime where it is exquisitely sensitive to affinity or quite insensitive.

Valency effects

As detailed in the introduction, ligand valency has been shown to affect the outcome of the sorting process in several systems. Our model predicts that monovalent and bivalent ligands may affect the sorting outcome very differently for many values of the parameters. To show this, the results of the model for case I and case II must be compared. One such comparison can be made using Figure 6.6. If we consider the situation in which only the transport of crosslinked receptors into the tubule is hindered, then $\gamma 1 = 1$ and $\gamma 2 < \gamma 1$. The kinetics of receptor accumulation in the monovalent and bivalent ligand cases can be considered by comparing the top curve ($\gamma 1 = 1$) in Figure 6.6a and one of the lower curves ($\gamma 1 = 1$ and $\gamma 2 < \gamma 1$) in Figure 6.6b. For the same sorting time, many more receptors will be recycled when the ligand is monovalent than when the ligand is bivalent.

Effect of tubule and ligand sizes

Changes in the size of the ligand molecule itself or in the diameter of the tubule alter the value of the partition coefficient κ. These effects are shown in Figures 6.9a and 6.9b for monovalent and bivalent ligands, respectively. In each plot, a decrease in the value of κ, meaning that the ligand appears larger relative to the tubule dimension, forces more of the free ligand to stay in the vesicle, binding to receptors there and not allowing them to enter the

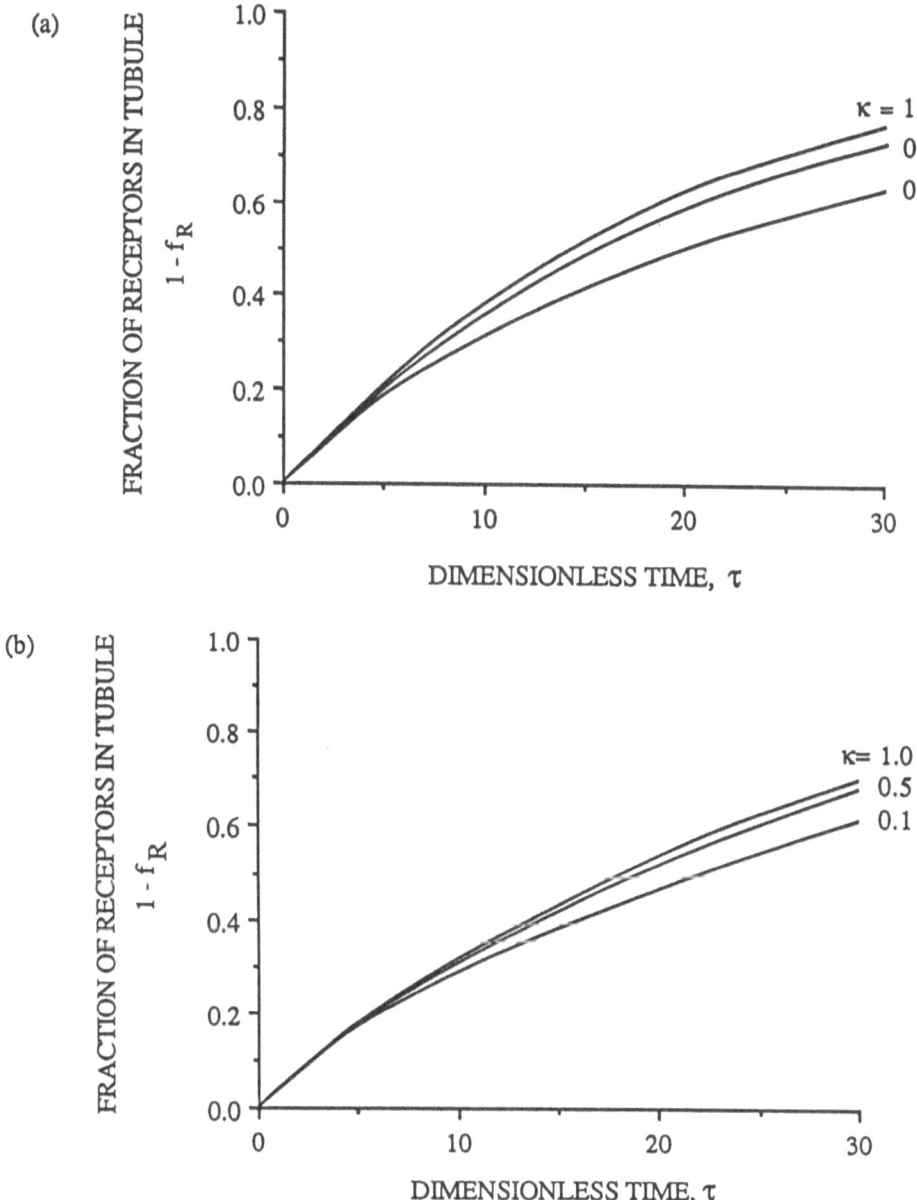

FIGURE 6.9 Effect of partition coefficient κ on receptor recycling. (a) Monovalent receptor, monovalent ligand system. Parameter values: $\varepsilon = 0.1$, $\eta = 100.0$, $\rho = 1.001$, $\gamma 1 = 0$, $f_b = 1.0$. (b) Monovalent receptor, bivalent ligand system. Parameter values: $\varepsilon = 0.1$, $\eta = 100.0$, $\chi = 100.0$, $\sigma = 1.0$, $\rho = 0.51$, $\gamma 1 = 0.0$, $\gamma 2 = 0.0$, $\overline{R}_f(0) = 0.0$, $\overline{L}_1(0) = 0.01$, $\overline{L}_2(0) = 0.99$.

tubules. Note that in Figure 6.9b both the transport parameters $\gamma 1$ and $\gamma 2$ are set equal to zero; if $\gamma 1$ is instead equal to one the effect of a change in κ is even less than that shown. For many values of the other relevant parameters, the value of κ has little influence on the fraction of receptors recycled. At very low values of κ our assumption of instantaneous equilibration of the ligand between the vesicular and tubular volumes of the endosome is no longer valid. In this range, we would also need to include the ligand transport kinetics in our model.

6.4 Comparison to experiment

General comparisons to monovalent receptor, monovalent ligand systems

The experimental observations noted in the introduction can all be accounted for by the model, which uses the parameters and initial conditions noted, for convenience, in Table 6.2. We begin with systems of monovalent receptors bound by monovalent ligands. If we assume first that ligand binding has no effect on receptor transport, or that $\gamma 1$ is equal to one, then receptors will be efficiently recycled if the sorting time is long enough. Many observations on differing amounts of ligand exocytosis in these systems can now be explained by suggesting a variation only in the binding rate constant k_1 and/or the dissociation rate constant k_{-1}, measurable parameters that will affect the values of the dimensionless groups ε and η. Several possibilities are shown in Figure 6.10a, where we plot the effect of a change in k_{-1} through the parameters ε and η when $\gamma 1 = 1$. In the transferrin receptor system, for example, the efficient return of ligand to the cell surface can be explained by a very low value of k_{-1} at the endosome pH so that ε is small and η is large (curve a). A low value of k_{-1} has indeed been measured for transferrin and its receptor in K562 cells (Klausner et al. 1983a). For this system, then, the explanation of large amounts of ligand exocytosis is simple: no significant fraction of the ligand ever dissociates from its receptor during passage through the cell (Klausner et al. 1983b). Systems with greater values of k_{-1} (curves c-e) may include the Fcγ receptor with a ligand formed by the Fab fragment of an anti-receptor antibody (Mellman et al. 1984) and the asialoglycoprotein receptor with the asialotriantennary glycopeptide ligand (Townsend et al. 1984). In these systems, intermediate amounts of ligand exocytosis, on the order of 50% to 75% of the total amount of ligand internalized, have been measured. Though the value of k_{-1} may be greater than in the transferrin system, a small although still significant value of the equilibrium parameter η means that an appreciable amount of the ligand will remain bound to receptors in the tubules of the endosome. Finally, there may be systems in which k_{-1} is very large or k_1 is very small, such that η is therefore very small (curve g). In these systems, then, there would be little ligand exocytosis. The insulin receptor on adipocytes and its ligand insulin may fall into this category, because Marshall (1985a) found only 25% ligand

(a)

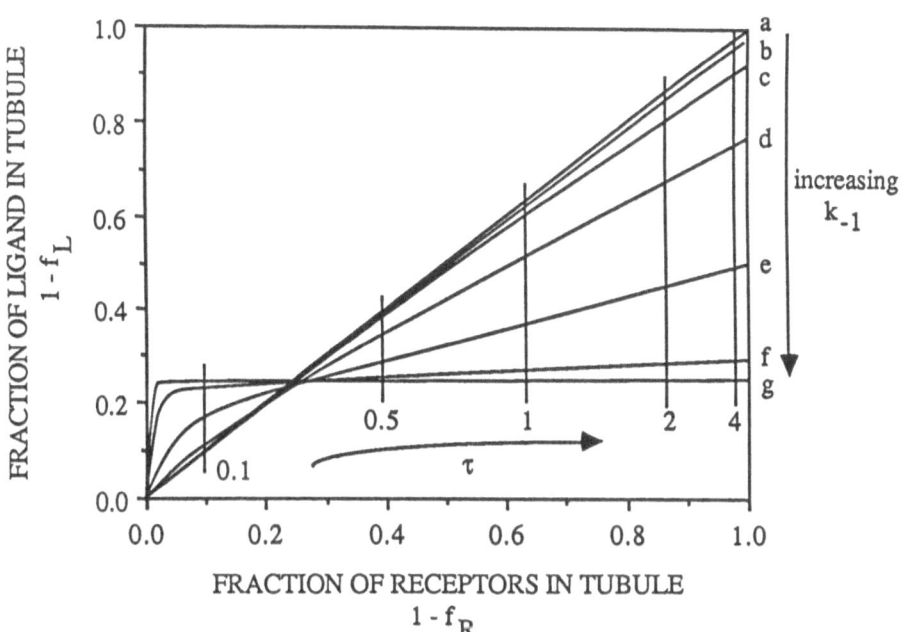

FIGURE 6.10 Effect of changing dissociation rate constant k_{-1} for monovalent receptor, monovalent ligand systems. For all plots, $\rho = 1.001$, $\kappa = 0.75$, $f_b = 1.0$, $V_y/V_T = 7/3$. The ε and η values are: curve a: $\varepsilon = 0.001$, $\eta = 10000$; curve b: $\varepsilon = 0.01$, $\eta = 1000$; curve c: $\varepsilon = 0.1$, $\eta = 100$; curve d: $\varepsilon = 1$, $\eta = 10$; curve e: $\varepsilon = 10$, $\eta = 1$; curve f: $\varepsilon = 100$, $\eta = 0.1$; curve g: $\varepsilon = 1000$, $\eta = 0.01$. (a) $\gamma 1 = 1$. (b) and (c) $\gamma 1 = 0$.

(b)

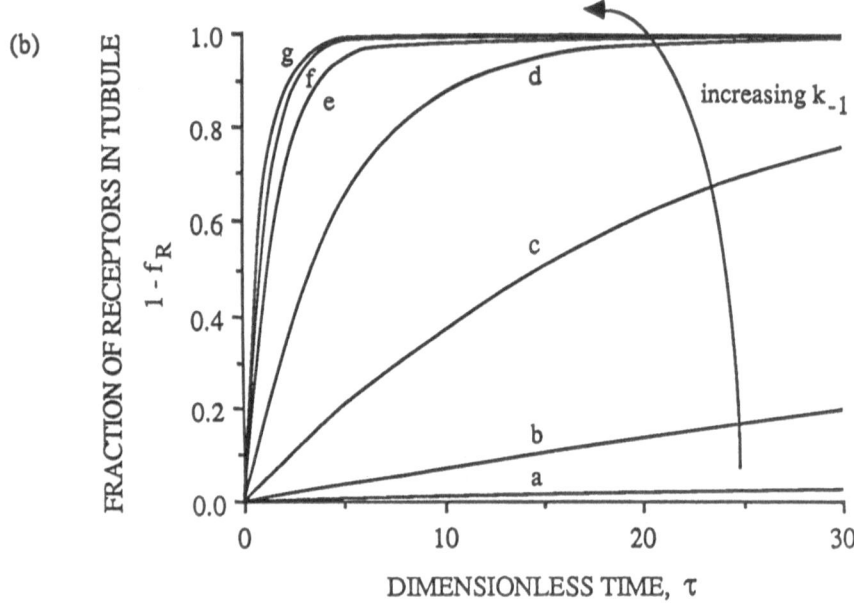

(c)

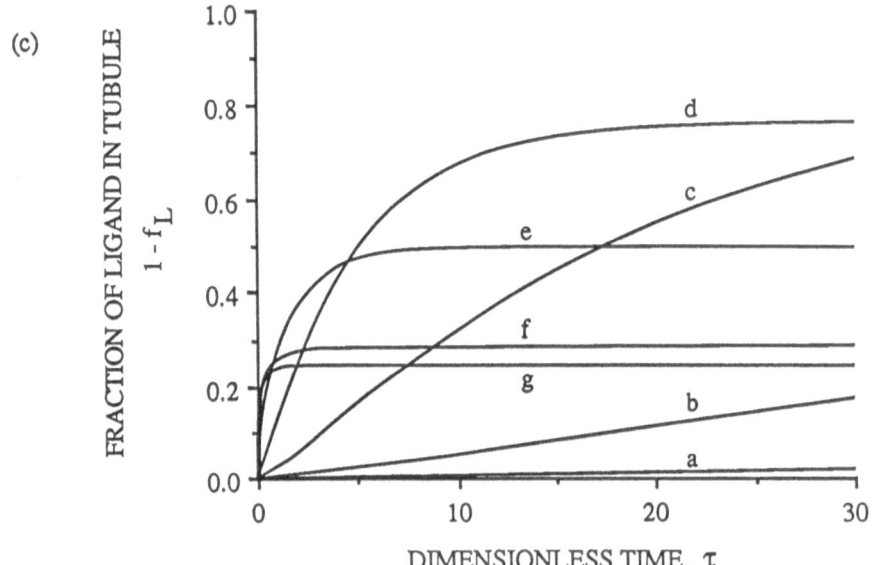

exocytosis.

There may also be monovalent receptor, monovalent ligand systems in which $\gamma 1$ is less than one. In many cells, neither the EGF receptor nor its ligand are returned to the surface in significant amounts (Stoscheck and Carpenter 1984). The receptors and ligand are either degraded or delivered to another intracellular location (Basu et al. 1985). The affinity of the ligand for the receptor may be great enough that few ligand molecules dissociate in the endosome, and if $\gamma 1$ is small most receptors and ligands will not recycle.

In Figures 6.10b and 6.10c, we plot the fractions of receptors and ligand recycled for the same parameter values as in Figure 6.10a but with the transport parameter $\gamma 1$ equal to zero. The fraction of receptors recycled increases as the dissociation rate constant k_{-1} is increased. The effects on ligand recycling are more complex. At low values of k_{-1}, an increase in k_{-1} allows more receptors to move into the tubule. Yet the affinity is great enough that many of those receptors again bind ligand once inside the tubule. The amount of ligand in the tubules can thus be great, because both bound and free ligand are present in substantial quantities. As the dissociation rate constant is increased, the affinity of the ligand for the receptors becomes small and essentially only free ligand is found in tubules, so that there is less total ligand in the tubules.

General comparisons to monovalent receptor, bivalent ligand systems

We can next examine experimental data on the sorting of monovalent receptor, bivalent ligand systems. The observations noted in the introduction can be accounted for by our model if we assume that $\gamma 1$ is close to one and $\gamma 2$ less than $\gamma 1$. As shown in Figure 6.8c, if the affinity of the bivalent ligand for the receptor is great enough, few receptors will recycle in the allotted sorting time. This may be the case when the transferrin receptor or the asialoglycoprotein receptors are bound by anti-receptor antibodies (Hopkins and Trowbridge 1983; Weissman et al. 1986; Schwartz et al. 1986) . On the other hand, if the affinity of the bivalent ligand for the receptor at low pH is not great (or if $\gamma 2$ is equal to one), receptor recycling will not be affected. The binding of the LDL receptor by the monoclonal antibody of Anderson et al. (1982) may fall into this category.

Extrapolation of model to more complex systems

Finally, there are more complex cases than those analyzed in this chapter for which our modeling may offer insight. For example, the binding of immune complexes or synthetic multivalent ligands to the Fc_γ receptors on macrophages is a situation in which the ligand valency may be greater than two. Yet we again predict that if the transport parameters dictate that aggregated receptors cannot move easily into tubules and the crosslinking rate constants of the ligands are such that much of the ligand stays bound to its receptor during sorting, these ligands will inhibit receptor recycling. This decrease in receptor recycling has been observed (Mellman and Plutner 1984; Ukkonen et al. 1986; Mellman et al. 1983), as detailed in Chapter 5.

Similarly, the binding of asialoorosomucoid to the asialoglycoprotein receptor on hepatocytes is probably a multivalent attachment. The observation of Townsend et al. (1984) that some of this ligand is exocytosed still bound to its receptor can be explained if the rate constants are such that at the low endosome pH a significant amount of the receptors in tubules are bound by ligand.

As a third example, Anderson et al. (1982) found that LDL receptors bound by a polyclonal antibody did not recycle to a significant extent, but that receptors bound by a monoclonal antibody did recycle. We suggest that the aggregated receptors in this system are unable to enter tubules easily ($\gamma2 < 1$) and that the extensive crosslinking induced by a polyclonal antibody is unlikely to be broken up in the short sorting time allowed by a cell. That the receptors internalized with the monoclonal antibody recycle suggests that the monoclonal antibody is less likely to maintain crosslinks throughout sorting; indeed, Goldstein and co-workers (Goldstein et al. 1982) report that the polyclonal antibody is less likely to dissociate from receptors than the monoclonal antibody.

Mellman et al. (1984) have also found that if Fc receptors are bound by the monovalent Fab ligand and then crosslinked prior to internalization with an antibody to that ligand, F(ab')$_2$, more internalized ligand is routed to lysosomes than if only the monovalent ligand alone is used. The crosslinking of receptors via the double ligand combination, Fab and F(ab')$_2$, can be predicted to decrease receptor recycling if $\gamma2$ is less than one and a substantial fraction of the crosslinks survive the low pH treatment in the endosome.

The model also extrapolates well to the IgE receptor system mentioned in the introduction and shown in Figure 6.1 (Furuichi et al. 1986). In this system, the binding of IgE to receptors has been shown to be relatively insensitive to a lowering of the pH to below 5.0 (Isersky et al. 1979), so the receptor and its bound ligand may be considered as a single bivalent receptor. Receptors bound by IgE but not crosslinked by antigen binding to the IgE molecule were found to recycle, as would be predicted by the model if $\gamma1 \approx 1$. Receptors bound by IgE and crosslinked by antigen did not recycle to a significant extent, suggesting that the transport parameter describing the movement of these crosslinked bivalent receptor, bivalent ligand complexes into tubules is less than one. In another experiment, receptors bound by IgE and crosslinked by antigen were exposed to a monovalent antigen that competed for the same binding site on the IgE molecule as the bivalent antigen. If the monovalent antigen was added to the cells before internalization or shortly after internalization, many of the receptors recycled. This suggests that the monovalent antigen caused the release of many of the crosslinks and thus the movement of many of the receptors into endosome tubules, as would be predicted if binding and dissociation rate constants were such that the monovalent ligand could compete effectively with the bivalent ligand within the sorting time and if $\gamma 1 \approx 1$.

Effect of raising the endosome pH

Chloroquine, NH_4Cl, and other acidotropic agents which act to raise the pH of intracellular compartments such as endosomes and lysosomes have been shown to affect the sorting process. The consequences of a raised endosome pH on receptor recycling can be quantitatively predicted by our model. If the transport parameters ($\gamma 1$ in a monovalent ligand system or both $\gamma 1$ and $\gamma 2$ in a bivalent ligand system) are equal to one, then no alteration in receptor recycling is expected. For example, chloroquine was not found to affect insulin receptor recycling in rat adipocytes (Ezaki et al. 1986). However, if a ligand has a greater dissociation rate at the normal low endosome pH that at a higher pH and if the transport parameters $\gamma 1$ and $\gamma 2$ specify that bound and/or crosslinked receptors cannot enter tubules, then receptor recycling will decrease if the endosome pH is raised. Quantitatively, this effect can be seen in Figure 6.10b for a monovalent receptor, monovalent ligand system because raising the endosome pH in a system in which binding is pH-sensitive is equivalent to lowering the value of k_{-1}. Similar effects on receptor recycling can be predicted for bivalent and multivalent ligand systems. For example, incubation of I cells in medium containing chloroquine led to a time-dependent and ligand concentration-dependent decrease in mannose 6-phosphate surface receptor number when those receptors were allowed to internalize multivalent ligands (Gonzalez-Noriega et al. 1980).

Ligand exocytosis can also be altered by raising the endosome pH in pH-sensitive systems. In a monovalent receptor, monovalent ligand system with $\gamma 1$ equal to one, raising the pH and therefore lowering k_{-1} will result in a greater amount of ligand exocytosis (Figure 6.10a). In accordance with this prediction, Greenspan and St. Clair (1984) measured low density lipoprotein exocytosis using monkey fibroblasts and found that exocytosis increased in the presence of chloroquine and methylamine. Predictions can also be made for Case 1 systems with $\gamma 1 < 1$ (for example, see Figure 6.10c) and for multivalent ligand systems.

Brown et al. (1986) found that the majority of the mannose-6-phosphate receptors on Clone 9 hepatocytes are intracellular and involved in the transport of newly synthesized multivalent lysosomal enzymes, an endogenous ligand, from the Golgi complex to endosomes. The remaining 10% of receptors are found on the cell surface and mediate the delivery of extracellular ligand to endosomes. The receptors in the endosome are then sorted from their ligands, and most receptors are recycled to the Golgi. As in many other receptor systems, the recycling of receptors is inhibited by NH_4Cl treatment. This can be explained by our model, as detailed above, by assuming that crosslinked receptors cannot enter tubules easily ($\gamma 2 < 1$) and that ligand binding is sensitive enough to low pH that at the low pH of the endosome most lysosomal enzymes dissociate within the sorting time. Brown and co-workers also found that when the monovalent ligand mannose 6-phosphate was added to the medium in the presence of NH_4Cl that receptor recycling could be induced. This effect was dependent on the concentration of the monovalent ligand; at low concentrations recycling did not occur but at high concentrations recycling resumed. Our model also predicts this result. As stated by Brown and co-workers, if the monovalent ligand appears in the same endosomes as the receptors crosslinked by the multivalent ligand, this new ligand will aid in the dissociation of the multivalent ligand. Clearly this effect will be dependent on the concentration of the monovalent ligand because enough ligand must be internalized to be effective as a competitive inhibitor of lysosomal enzyme binding. Thus at high concentrations of the monovalent ligand, few receptors will be crosslinked. If the transport parameter $\gamma 1$ is equal to one so that bound but uncrosslinked receptors are not hindered in their entry into the tubules, then the model predicts that receptor recycling will occur.

We also note that the acidotropic agents may not only alter the pH of the endosome (and therefore sorting) and lysosome but may have other effects on the endocytic cycle as well, as described by Mellman et al. (1986).

Quantitative comparison to the chemotactic peptide receptor system

The chemotactic peptide receptor system of polymorphonuclear leukocytes is a monovalent receptor, monovalent ligand system in which the transport parameter $\gamma 1$ is likely less than one. The extracellular ligand concentration L_0 has been shown to affect receptor recycling in the chemotactic peptide receptor system; Zigmond et al. (1982) report that as the extracellular concentration of a chemotactic peptide FNLLP (N-formylnorleucylleucylphenyl-alanine) was increased from 10^{-8} M to 2×10^{-7} M, the rate constant for receptor recycling decreased from 0.11 min^{-1} to 0.038 min^{-1} (by a whole cell kinetic analysis similar to ours). To account for this, we assume again that the vesicular transport rate constants, k_R and k_L, and the sorting time do not change, or that only the sorting fractions f_R and f_L are altered by an increase in the extracellular ligand concentration. According to our model, for the value of L_0 to affect the receptor sorting fraction f_R, $\gamma 1$ must be less than one.

An approximate comparison of model predictions with experimental data can be performed as follows. The ratio of the experimental value of the sorting fraction $1 - f_R$ at either of two ligand concentrations, 3×10^{-8} and 2×10^{-7} M, to the value at the lowest ligand concentration, 10^{-8} M, was found by taking ratios of the experimentally determined recycling rate constant $k_R((1-f_R)$ at the appropriate concentrations. This information is given in Table 6.3.

The values of the association and dissociation rate constants $k_1{}^N$ and $k_{-1}{}^N$ are reported as 2×10^7 $M^{-1}min^{-1}$ and 0.4 min^{-1} respectively. These values are not significantly effected by lowering the pH to 4.0 (Zigmond et al. 1982), so these are also good estimates for k_1 and k_{-1}, the association and dissociation rate constants inside the endosome. The volume of the endosome was assumed to be equal to 4×10^{-14} cm^3, 70% of which is in the central vesicle (Marsh et al. 1986). The FNLLP ligand is small, so the partition coefficient κ is estimated at about 0.9. The rate constant k_a will be assumed to be on the order of 1 min^{-1}, as detailed earlier in this chapter. Thus ϵ is equal to 0.4.

We also assume that the free and bound receptors on the cell surface are endocytosed in proportion to their existence on the cell surface at equilibrium. Thus the initial condition, the parameter f_b, depends on the ligand concentration and can be calculated from Eq. 6.19. The total number of receptors in each endosome, needed to calculate the values of η and ρ, is unknown and will be estimated at 500. The model predictions are not particularly sensitive to this value (data not shown). Finally, the transport parameter $\gamma 1$ was set equal to zero.

Using these parameter estimates and our model, the fraction of receptors recycled at each of the three ligand concentrations can be calculated. These fractions are plotted in Figure 6.11 as a function of the dimensionless time. In Table 6.3, the predicted ratios of the recycled fractions at two ligand concentrations are given for comparison with the experimental values. Note that for dimensionless times τ in the range of 2 to 6, corresponding to sorting times of about 2 to 6 min., the model predictions match the experimental sorting fractions extremely well.

Quantitative comparison to the Fc_γ receptor system

A comparison of model predictions to data on the Fc_γ receptor system on J774 cells (Mellman et al. 1984; Mellman and Plutner 1984; Ukkonen et al. 1986) is more difficult than the comparison to the chemotactic peptide receptor system. As detailed in Chapter 5, the most useful comparison to make is one between the monovalent and multivalent ligand systems, and this requires the estimation of many more parameters. In addition, the valency of the multivalent ligand used, a ligand formed by the adsorption of Fab onto colloidal gold, is not known. Ukkonen et al. (1986) estimate that between 2 and 12 Fab ligands are adsorbed onto each gold particle, and it is not known what fraction of these binding sites are simultaneously accessible to the receptors. For the purposes of a crude comparison with the experimental data, it will be assumed that this multivalent ligand is functionally bivalent.

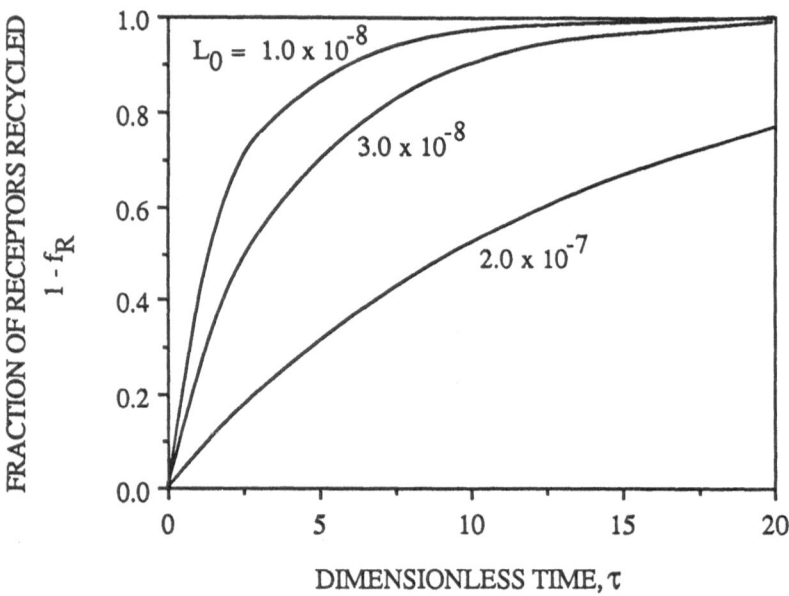

FIGURE 6.11 Single endosome model predictions for the chemotactic peptide receptor system on polymorphonuclear leukocytes. The fraction of receptors recycled as a function of time and extracellular ligand concentration is shown for the parameter values given in the legend of Table 6.3 and in the text.

ligand concentration (M)	$k_R(1-f_R)^{exp}$ (min^{-1})	$\left[\dfrac{(1-f_R)^{(i)}}{(1-f_R)^{(1)}}\right]^{exp}$	$(1-f_R)^{theor}$ (τ)	$\left[\dfrac{(1-f_R)^{(i)}}{(1-f_R)^{(1)}}\right]^{theor}$ (τ)
(1) 1×10^{-8}	0.11	1.0	0.63 (2) 0.81 (4) 0.90 (6)	1.0 (2) 1.0 (4) 1.0 (6)
(2) 3×10^{-8}	0.086	0.78	0.43 (2) 0.63 (4) 0.76 (6)	0.68 (2) 0.78 (4) 0.85 (6)
(3) 2×10^{-7}	0.038	0.35	0.14 (2) 0.26 (4) 0.36 (6)	0.23 (2) 0.32 (4) 0.40 (6)

TABLE 6.3 Experimental and predicted (theoretical) fractions of receptors recycled for the chemotactic peptide receptor system on polymorphonuclear leukocytes. Parameters values found according to the text. For all ligand concentrations: $\epsilon = 0.4$, $\eta = 1.07 \times 10^3$, $\kappa = 0.9$, $\gamma_1 = 0$. For $L_0 = 1 \times 10^{-8}$, $f_b = 0.33$, and $\rho = 0.330$. For $L_0 = 3 \times 10^{-8}$, $f_b = 0.60$, and $\rho = 0.601$. For $L_0 = 2 \times 10^{-7}$, $f_b = 0.91$, and $\rho = 0.916$. $i = 1, 2, 3$.

The values of several parameters can be estimated from experimental data reported by Mellman and co-workers (Mellman et al. 1984; Mellman and Unkeless 1980): $k_1N = 3 \times 10^6$ $M^{-1}min^{-1}$, $k_{-1}N = 0.0023$ min^{-1}, $k_{-1} = 0.1$ min^{-1}, and, assuming that only the dissociation rate constant is affected by low pH, $k_1 = 3 \times 10^6$ $M^{-1}min^{-1}$. These values are appropriate for both the monovalent ligand and the "bivalent" ligand. The ligand concentration can be calculated to be approximately 2×10^{-8} M for the monovalent ligand and 10^{-8} M for the multivalent ligand. The volume of the endosome is estimated at 4.0×10^{-14} cm^3, with 70% of that volume in the vesicle (Marsh et al. 1986)

We again estimate that k_a is on the order of 1 min^{-1}. The total number of receptors in the endosome was set to 1000, and the partition coefficient is estimated at 0.5 for the monovalent ligand and 0.25 for the "bivalent" ligand. The transport parameters will be set at $\gamma 1 = 1$ and $\gamma 1 = 0.5$, implying that crosslinked but not bound receptors are hindered in entering the tubules. We assume that only bound receptors enter the endosome and therefore f_b is equal to one in the case of the monovalent ligand. The crosslinking parameter χ is set equal to η and σ is set equal to one, in accordance with the general guidelines for parameter ranges described earlier in this chapter. Finally, crude estimates of the fraction of surface receptors crosslinked at equilibrium indicated that nearly all receptors are crosslinked and therefore $\overline{L_1}(0) \approx 0$ and $\overline{L_2}(0) \approx 1$.

Given these parameter estimates, all of the model parameters can be calculated. The model then predicts that the fraction of receptors degraded, f_R, will vary between the two types of ligand. This prediction is shown in Table 6.4 and compared with the result of the data fit performed in Chapter 5. Although the values of the parameters are only our best estimates, the model is able to predict the observed effect of ligand valency on the sorting process.

6.5   Discussion

In this chapter, we have proposed an explanation of the different outcomes of intracellular receptor/ligand sorting for various receptor systems by developing a mathematical model for the sorting process as it occurs in single endosomes. Our analysis of the sorting mechanism has involved a delineation of the reaction and transport kinetics occurring within the endosome together with several assumptions. We first assumed that receptors are initially located in the vesicular portion of the endosome and that receptors moving from the vesicle into tubule(s) within the sorting time are recycled to the cell surface. Most significantly, we then postulated that receptors moving into tubules are trapped there and that ligand properties affect receptor/ligand sorting through their effects on receptor transport. Taken together, these assumptions have enabled us to calculate the fractions of receptors and ligands which are returned to the cell surface or degraded as functions of the time allowed by the cell for sorting and the parameters describing the binding, dissociation, and transport events within an

dimensionless time τ	$(f_R^{monov})^{theor}$	$(f_R^{multiv})^{theor}$	$\left[\dfrac{f_R^{multiv}}{f_R^{monov}}\right]^{theor}$
1.0	0.37	0.59	1.6
2.0	0.14	0.35	2.5
4.0	0.02	0.12	6.0
6.0	0.002	0.04	20.

TABLE 6.4 Model predictions of the fraction of receptor degraded in the Fc$_\gamma$ receptor system on macrophages. Predictions should be compared with the experimental value of the ratio f_R^{multiv}/f_R^{monov}, 2.7, found in Chapter 5. Monovalent ligand parameter values: $\varepsilon = 0.1$, $\eta = 1500$, $\kappa = 0.5$, $\gamma 1 = 1.0$, $\rho = 1.0003$, $f_b = 1.0$. Multivalent ligand parameter values: $\varepsilon = 0.1$, $\eta = \chi = 1500$, $\kappa = 0.25$, $\gamma 1 = 1.0$, $\gamma 2 = 0.5$, $\rho = 0.50017$, $\sigma = 1.0$, $\overline{R}_f(0) = 0.0$, $\overline{L}_1(0) = 0.0$, $\overline{L}_2(0) = 1.0$.

endosome.

It is not known with certainty that the true direction of movement of receptors is from the vesicle to tubule(s). Alternatively, receptors might be located initially in tubules and move out of the tubules into the vesicle in order to gain entrance to the degradation pathway. This second alternative appears unlikely, however, when data on the effects of receptor crosslinking on receptor recycling are considered. In several systems (Mellman and Plutner 1984; Ukkonen et al. 1986; Hopkins and Trowbridge 1983; Weissman et al. 1986; Anderson et al. 1982; Townsend et al. 1984), ligands which can crosslink their receptors have been shown to inhibit receptor recycling. To explain this, one would need to postulate the existence of a positive signal, one which recognizes crosslinked receptors and enhances their movement out of tubules over the movement of the other receptors. We consider this to be less likely than the scheme suggested here in this model. In addition, it has been reported recently that in kidney proximal tubule cells the direction of movement of newly endocytosed material is from vesicles into tubules (Hatae et al. 1986). Finally, it is worthwhile to note that if tubules do grow from a vesicle, receptors moving from tubules to the central vesicle would need to diffuse against a membrane current moving in the opposite direction; it would seem more likely that receptors would move with the current, from the vesicle to a tubule.

We have assumed that the morphology of and pH inside the endosome do not change during the sorting process. There is recent evidence, however, that the central vesicle of the endosome may grow in size (Geuze et al. 1983; Geuze et al. 1987) and that the process of acidification is not instantaneous in relation to the sorting process (Kielian et al. 1986). These data indicate that the rate constants for binding, dissociation, crosslinking, and breakup of crosslinks, all of which may be pH-dependent, may therefore be time-dependent. The rate constant for the movement of receptors into tubules, because it is dependent on the geometry of the endosome (Eq. 6.52), would also be a function of time. These aspects can be included in the model once more information is available on the time course of acidification and changes in the endosome morphology as well as information on the variation of the required rate constants with pH.

The model presented here can be extended to the case of monovalent receptors with ligands of valency greater than two, to monovalent receptors with mixtures of monovalent and multivalent ligands and, with more difficulty, to systems of multivalent receptors and ligands. In the latter case, receptor aggregates are not limited in size and the mathematics become more complex (Macken and Perelson 1985). The model can also be used to predict the separation if receptors and ligands are initially distributed throughout the vesicle and tubule(s). In this case, the initial conditions for the equations would change, but the qualitative aspects of the model predictions would not vary significantly from those presented here.

This model suggests the need for the measurement or estimation of several fundamental parameters in order to make quantitative predictions for any particular receptor/ligand system. Specifically, values for the association and dissociation rate constants k_1 and k_{-1} and the two crosslinking rate constants k_2 and k_{-2} at the low pH of the endosome are required. One method of estimating k_2 and k_{-2} for symmetric bivalent ligands from binding data for the monovalent ligand with the same binding group is given in Appendix II. Measurements have been made of the vesicle and tubule dimensions and/or volumes in several cell types (Geuze et al. 1983; Marsh et al. 1986; Bretscher and Thomson 1985; Wall et al. 1980; Harding et al. 1985; Yamashiro and Maxfield 1984), but the values for the particular system in question should also be obtained. Estimates are also needed for the total number of receptors in an endosome and on the states of those receptors when they are delivered to the endosome for sorting. Finally, it would be helpful to have a direct measurement of the sorting time, although there are several estimates available.

If these fundamental parameters can be measured or estimated, then the remaining unknown model parameters include only the transport parameters γ_1 and γ_2. At this time, these parameters are free parameters which may be varied to fit the data; however, it is hoped that techniques to measure diffusion coefficients in the endosome will soon be developed. In many cases, it appears that $\gamma_1 \approx 1$ and $\gamma_2 \approx 0$ is a good fit to the available data. A recent experiment by Huecksteadt and co-workers (1986) on the insulin receptor in rat adipocytes, though not a direct measurement of the diffusion coeffient of bound receptors in endosomes, has shed light on the value of the transport parameter γ_1 in their system. An insulin analog was used to covalently label insulin receptors on the cell surface. These receptors were internalized and recycled with the same kinetics as unmodified insulin receptors, suggesting that $\gamma_1 \approx 1$.

This work also suggests the need for a systematic investigation into the variation of the sorting outcome with changes in the model parameters. Because the outcome of the sorting process cannot yet be measured directly, the model we have presented must be used in the context of a whole cell model for endocytosis. Whole cell data on endocytosis, including, for example, the changes in cell surface receptor number with time, initial rates of internalization, ligand exocytosis, receptor degradation, and receptor synthesis, can be obtained. These data can then be fit with the whole cell kinetic model to determine how the sorting fractions vary with ligand properties and experimental conditions. These experimentally observed variations in the sorting fractions can then be compared with the predictions of the single endosome model presented here.

In conclusion, it is hoped that the application of this model to many systems which exhibit receptor-mediated endocytosis will simplify the interpretation of the range of outcomes of the sorting process that have been observed.

CHAPTER 7
CONCLUSIONS

The intracellular sorting of receptors and ligands in the endosome is a critical step in the endocytic cycle because it is the point at which the cell targets molecules to their final destinations. By controlling the fraction of internalized receptors that recycle to the cell surface, the cell can regulate its ability to bind extracellular ligand and can thus modulate its response to a particular ligand. Similarly, a cell may choose to degrade ligands or to recycle them, a choice which again may influence the cell's ability to respond to a particular ligand. If this intracellular sorting process can be understood, it can perhaps be modified or exploited for biotechnological gain.

For example, nonresponsive cells might be made responsive by altering the fate of intracellular receptors or ligands. Analysis of the kinetic data of Schaudies and co-workers (Schaudies et al. 1985) on the endocytosis of epidermal growth factor (EGF) by various cell types has indicated that the ability of a cell to respond to EGF may be related to the outcome of the sorting process: the hypothesis that responsive cells recycle fewer of their internalized EGF receptors than nonresponsive cells is consistent with the experimental data for these cells (Lauffenburger et al. 1987). Using the model described here, one might be able to design a synthetic growth factor with a specific sorting outcome, for example, the outcome for maximal responsiveness.

As a second example, pharmaceuticals might be designed for maximum effectiveness, a term that for some drugs may mean maximal uptake while still allowing receptors to recycle and internalize additional ligand (drug) molecules. Or, if the drug acts only while bound to surface receptors and the internalization pathway serves as a "turn-off" signal, the drug might be designed to recycle while still bound to its receptors. Thus one also requires information on the mechanism of signal transduction, or the method by which a particular ligand is able to elicit a cellular response, and its dependence on the fates of intracellular receptors and ligands. In other words, one would need to know whether bound surface ligands, bound intracellular receptors, degraded receptors, and/or other receptor or ligand forms are responsible for the cellular response. This information would allow one to decide how the sorting outcome should be altered, an alteration that might be possible by modifying ligand properties such as affinity and valency according to the principles outlined here in our models. Thus the mechanism of intracellular receptor/ligand sorting is important to understand so that its outcome may be manipulated.

In this text, we have examined a variety of experimental data detailing the many possible outcomes of the sorting process. The objective of this study was to propose a sorting mechanism that could account for all of these sorting outcomes upon changes in fundamental and measurable parameters of the system. We began in Chapter 2 by presenting literature data on the sorting process that suggest that any proposed sorting mechanism must allow, for some values of the relevant parameters which control the outcome of sorting, for the return of at least 95% of all internalized receptors to the cell surface. In Chapter 3 we proposed and evaluated a simple sorting mechanism: the equilibration of dissociated receptors and ligands within the sorting organelle, the endosome, according to the distribution of surface area and volume, respectively, between the tubules and central vesicle of the endosome. It was found that too few receptors would accumulate in the presumed recycling vehicles, the tubules, to account for the experimental data.

In Chapter 4 we examined several possible kinetic sorting mechanisms. Sorting by a difference in capture times, the times required by receptors and ligands to move into endosome tubule(s) from the central vesicle, was found to be an unlikely mechanism given approximate receptor and ligand diffusion coefficients and the geometry of the endosome. Even in the presence of a convective membrane current of 1 to 3 μm/min acting to enhance receptor movement from the vesicle into a tubule, this mechanism cannot account for the observed separation of receptors and ligands. A second kinetic mechanism, the diffusion of receptors and ligands in the endosome with receptor trapping in tubules, was found to allow the rapid and efficient return of receptors to the cell surface, thus meeting the minimal requirements of a sorting mechanism proposed in Chapter 2.

This diffusion with trapping mechanism fails, however, to predict the substantial changes in the sorting outcome with changes in ligand properties that have been recently reported; the relevant data are detailed in Chapters 5 and 6. We therefore suggested a mechanism which includes more complex transport/reaction schemes. Significantly, we postulated that the state of a receptor (free, bound, or crosslinked) may affect the rate of receptor transport into tubules. This single endosome model, developed in Chapter 6, is able to predict a wide range of sorting outcomes that are dependent on the values of fundamental transport and reaction parameters of the system.

Most importantly, this model of the sorting process is testable. Our discussion of available experimental data in Chapters 5 and 6 showed that some of the data needed for a rigorous test of the model are currently available. In general, if ligands are constructed with known valencies and binding parameters and if data on internalization, degradation, recycling, and ligand exocytosis are obtained, the experimental sorting fractions can be calculated from the data according to the methods outlined in Chapter 5. These sorting fractions can be compared with the fractions predicted by the model, as described in Chapter 6. Thus the primary recommendation for future work on understanding the sorting process is to test the model proposed here experimentally.

In addition, several pieces of evidence suggest that it may be worthwhile to develop a continuous model of sorting based on the batch model presented here. There is evidence that the diameter of the endosome vesicle may change with time, increasing initially as more incoming endocytic vesicles fuse with the central vesicle of the endosome and possibly decreasing at later times as the vesicle matures and sorting draws to a close (Geuze et al. 1983; Geuze et al. 1987). Thus the transport parameters for receptors, which depend on the distance the receptor must travel to reach a tubule, may change with time. There are also data that suggest that receptors and ligands may be exposed to a continually decreasing pH on their journey from the cell surface to the endosome (Roederer et al. 1987). In our model, we assumed that acidification was instantaneous upon the delivery of molecules to the endosome and the start of the sorting time. Changes in pH with time mean that the reaction rate constants in our model are functions of time; in addition, the initial conditions for sorting will depend strongly on the environment the molecules experience during transport from the surface to the endosome. Finally, tubules may possibly form from or fuse with the central vesicle and then disconnect from the central vesicle at several times during the sorting process. Quantitative data on all of these changes in geometry and pH are not currently available and would be important in the development of a continuous sorting model. In addition, information on the pH dependence of the binding and dissociation rate constants would be necessary. It should be clear from Table 6.1 that this rate constant data are not currently available for nearly all receptor/ligand systems.

Our model has assumed that free ligand molecules are simply partitioned between the vesicle and tubule volumes. We note that this partitioning may be complicated considerably by the presence of attractive or repulsive electrostatic forces which may act to encourage or discourage the accumulation of ligand in the tubules.

An important question that remains to be addressed is the mechanism of formation of the endosomes themselves. Do tubules attach to large vesicles formed by the fusion of several smaller endocytic vesicles? Or do tubules grow from the endosome, and, if so, what factors influence the growth? The formation of the tubules may have a role in sorting itself by causing a membrane current to draw receptors into tubules, as discussed in Chapter 4. As a third alternative, we suggest that the tubules may form as the vesicle itself forms. If the volume of fluid inside primary endocytic vesicles remains constant when these vesicles fuse to form the endosome, and if we assume that the surface area of the fusing vesicles does not change (Evans and Skalak 1980), this constant volume requirement implies that the vesicle resulting from the fusion will no longer be spherical. The extra membrane might be used to form the low volume, high surface area tubule.

The experimental literature shows that several receptor/ligand systems can be sorted simultaneously, and the sorting model presented here can be applied independently to any number of systems. A more complicated situation is that in which molecules from different receptor/ligand systems are targeted to more than two locations simultaneously. For example, asialoglycoprotein (ASGP) receptors, mannose-6-phosphate (M6P) receptors, and polymeric immunoglobulin A (IgA) receptors have been found together in the endosomes of rat hepatocytes (Geuze et al. 1984). The IgA receptor, which internalizes its ligand on one side of the cell and is then recycled together with its ligand to the opposite side of the cell, was found in tubules, suggesting that it was about to be recycled. The ASGP and M6P receptors were found in different tubules or tubule microdomains than those in which the IgA receptor was found. This suggests that the "trapping mechanism" found in the tubules may in fact have some degree of molecular specificity. This is not unexpected, for the coated pits which trap molecules on the cell surface are known to concentrate some molecules but not others (Wileman et al. 1985a).

Finally, the sorting ideas presented here may have implications for sorting along other pathways of intracellular traffic. For example, protein secretion from cells can occur by at least two routes. Proteins secreted by the constitutive pathway are secreted into the environment at low concentration at the same rate as they are synthesized; proteins secreted by the regulated pathway are stored at high concentrations in specialized vesicles and released only when the cell is appropriately stimulated. Proteins secreted by both pathways are made concurrently in the same intracellular location and therefore must be separated prior to release. It is thought that this separation is at least in part mediated by receptors, and there is evidence for a low pH step in the process and also for "mis-sorting" of proteins (Kelly 1985; Pfeffer and Rothman 1987). One might apply the general principles here to predict the fraction of proteins secreted immediately or stored as a function of fundamental parameters such as binding and dissociation rate constants, geometry, pH, and transport rate constants.

APPENDIX I
DERIVATION OF THE GREEN'S FUNCTION

The Green's function for Laplace's equation in a sphere with a constant flux boundary condition over the entire surface of the sphere can be found in the following way. The solution to Laplace's equation is assumed to be the product of separable functions of the three variables r, θ, and ϕ. The relevant eigenvalue problems are then solved and the requirements of a Green's function are used to solve for any unknown constants.

The angular portion of the kernel can be found by these techniques and is given by Weinberger (1965) as

$$K_n(\theta,\phi \,;\, \theta',\phi') = \frac{2n+1}{4\pi} P_n(\cos\gamma), \qquad (A1.1)$$

where

$$\cos\gamma = \cos\theta\cos\theta' + \sin\theta\sin\theta'\cos(\phi-\phi') \qquad (A1.2)$$

and P_n are the Legendre polynomials of degree n. From the associated eigenvalue problem it is known that $\lambda^2 = n(n+1)$. The remaining part of the problem is to find the radial part of the Green's function, $J_n(r \,;\, r')$. The full Green's function is then found from

$$G(r, \theta, \phi \,;\, r', \theta, \phi) = \sum_{n=0}^{\infty} J_n(r \,;\, r') K_n(\theta, \phi \,;\, \theta', \phi'). \qquad (A1.3)$$

To find the radial portion of the Green's function, we must solve

$$\frac{\partial}{\partial r}\left(r^2 \frac{\partial J_n}{\partial r}\right) - \lambda^2 J_n = 0 \qquad (A1.4)$$

to find that

$$J_n(r) = A_n r^n + B_n r^{-(n+1)}. \tag{A1.5}$$

The next step is to find the unknown constants A_n and B_n by specifying that the $J_n(r)$ meet certain requirements of a Green's function. To begin, we define

$$J_n^1(r) = A_n^1 r^n + B_n^1 r^{-(n+1)} \qquad 0 < r < r' \tag{A1.6}$$

$$J_n^2(r) = A_n^2 r^n + B_n^2 r^{-(n+1)} \qquad r' < r < R, \tag{A1.7}$$

where the superscripts 1 and 2 on J refer to region one, $0 < r < r'$, and region two, $r' < r < R$, respectively.

Take first the case $n \neq 0$. We require that for $0 < r < r'$ the function $J_n(r)$ remains bounded at $r = 0$, and therefore $B_n^1 = 0$. For $r' < r < R$, we set for simplicity $\partial J_n^2(r)/\partial r$ equal to zero at $r = R$. It is also necessary to satisfy a continuity condition, that the value of J_n^1 as r approaches r' from the left is equal to the value of J_n^2 as r approaches r' from the right. In addition, the condition of a jump discontinuity in the derivative must be satisfied, so that

$$\left. \frac{dJ_n^2}{dr} \right|_{r=r'} - \left. \frac{dJ_n^1}{dr} \right|_{r=r'} = \frac{-1}{r'^2}. \tag{A1.8}$$

Taken together, these requirements allow the remaining three unknowns A_n^1, A_n^2, and B_n^2 to be determined. After algebraic simplification, one finds the symmetric functions

$$J_n^1(r) = \frac{1}{(2n+1)R} \left(\frac{r}{R}\right)^n \left[\left(\frac{n+1}{n}\right)\left(\frac{r'}{R}\right)^n + \left(\frac{r'}{R}\right)^{-(n+1)} \right] \tag{A1.9}$$

$$J_n^2(r) = \frac{1}{(2n+1)R} \left(\frac{r'}{R}\right)^n \left[\left(\frac{n+1}{n}\right)\left(\frac{r}{R}\right)^n + \left(\frac{r}{R}\right)^{-(n+1)} \right]. \tag{A1.10}$$

One must also consider the case of $n = 0$. As for the case for $n \neq 0$, we define

$$J_0^1(r) = A_0^1 + B_0^1 r^{-1} \qquad 0 < r < r' \tag{A1.11}$$

$$J_0^2(r) = A_0^2 + B_0^2 r^{-1} \qquad r' < r < R \tag{A1.12}$$

We then apply the conditions as for $n \neq 0$ of boundedness at $r = 0$, continuity at $r = r'$, and a jump discontinuity in the derivative at $r = r'$. We do not require, however, that $\partial J_0^2(r)/\partial r$ is equal to zero at $r = R$, because the overall derivative of the Green's function with respect to r must not be equal to zero at the outer boundary of the sphere, $r = R$, if we are to use the Green's function to satisfy a problem with a nonzero derivative over some part of the surface of the sphere. Thus

$$J_0^1(r) = A_0^2 + \frac{1}{r'} \qquad\qquad 0 < r < r' \qquad\qquad (A1.13)$$

$$J_0^2(r) = A_0^2 + \frac{1}{r} \qquad\qquad r' < r < R. \qquad\qquad (A1.14)$$

Note that there remains still one undetermined parameter, A_0^2. This we will choose to be equal to $1/R$ so that the simplification of the Green's function, as detailed in the next few equations is straightforward. Our choice of A_0^2 will then force the unknown constant C in the problem formulation of Chapter 5 to take on a particular value in order to satisfy Fredholm's Alternative, the requirement of solvability (see Eq. 4.19).

Using this value A_0^2 and summing over all n,

$$G = \frac{1}{4\pi} \sum_{n=0}^{\infty} \frac{1}{R} \left(\frac{r\,r'}{R^2}\right)^n P_n(\cos\gamma) + \frac{1}{4\pi} \sum_{n=1}^{\infty} \left(\frac{1}{R}\right)\left(\frac{1}{n}\right)\left(\frac{r\,r'}{R^2}\right)^n P_n(\cos\gamma)$$

$$+ \frac{1}{4\pi} \sum_{n=0}^{\infty} r^n \, r'^{-(n+1)} P_n(\cos\gamma) \qquad\qquad 0 < r < r' \qquad\qquad (A1.15)$$

and

$$G = \frac{1}{4\pi} \sum_{n=0}^{\infty} \frac{1}{R} \left(\frac{r\,r'}{R^2}\right)^n P_n(\cos\gamma) + \frac{1}{4\pi} \sum_{n=1}^{\infty} \left(\frac{1}{R}\right)\left(\frac{1}{n}\right)\left(\frac{r\,r'}{R^2}\right)^n P_n(\cos\gamma)$$

$$+ \frac{1}{4\pi} \sum_{n=0}^{\infty} r'^n \, r^{-(n+1)} P_n(\cos\gamma) \qquad\qquad r' < r < R. \qquad\qquad (A1.16)$$

This Green's function is simplified by the application of the two identities (Weinberger

1965; Koshlyakov et al. 1964):

$$\frac{1}{b} \sum_{n=0}^{\infty} (\frac{a}{b})^n P_n (\cos \gamma) = (a^2 + b^2 - 2ab \cos \gamma)^{-1/2} \qquad a < b \qquad \text{(A1.17)}$$

$$\sum_{n=1}^{\infty} (\frac{P_n (\cos \gamma)}{n}) (\frac{cb}{a^2})^n = \ln \left[\frac{2a^2}{a^2 - cb \cos \gamma + (a^4 + c^2 b^2 - 2a^2 cb \cos \gamma)^{1/2}} \right]. \qquad \text{(A1.18)}$$

The result is the desired Green's function

$$G(r,\theta,\phi ; r',\theta',\phi') = \frac{1}{4\pi} \left[\frac{R}{(R^4 + r^2 r'^2 - 2rr'R^2 \cos \gamma)^{1/2}} \right.$$

$$+ \frac{1}{R} \ln \left(\frac{2R^2}{R^2 - rr' \cos \gamma + (R^4 + r^2 r'^2 - 2rr'R^2 \cos \gamma)^{1/2}} \right)$$

$$\left. + (r^2 + r'^2 - 2rr' \cos \gamma)^{-1/2} \right] \qquad \text{(A1.19)}$$

APPENDIX 2
ESTIMATION OF THE RATE CONSTANTS k_1, k_{-1}, k_2 AND k_{-2}

If the association and dissociation rate constants for the binding of a monovalent ligand to a cell expressing surface receptors for that ligand have been measured, the association and dissociation rate constants per receptor can be determined. In addition, the binding and crosslinking rate constants for a symmetric, bivalent ligand with two copies of the same receptor-binding site as the first ligand can be estimated. The binding of these two ligands are shown schematically in Fig. A2.1.

Monovalent receptor with monovalent ligand

The measured association and dissociation rate constants for the binding of a monovalent ligand to a cell are defined as $k_1{}^C$ and $k_{-1}{}^C$. It is important to note that these are per cell, not per receptor, rate constants. The overall rate constants $k_1{}^C$ and $k_{-1}{}^C$ are composed of three parts: the transport rate constants for the movement of ligand to the cell surface, the transport rate constants for the formation of a receptor/ligand encounter complex at the cell surface, and the association and dissociation rate constants reflecting the intrinsic affinity of the ligand for the receptor. This can be represented as:

$$ R \; + \; L \underset{k_-^S}{\overset{k_+^S}{\rightleftarrows}} R \; + \; L^{surf} \underset{k_-'}{\overset{k_+'}{\rightleftarrows}} R\!-\!L \underset{k_{-i}}{\overset{k_i}{\rightleftarrows}} RL $$

where R represents a free receptor, L a free ligand molecule in solution, and L^{surf} a free ligand molecule at the cell surface. R—L is termed an encounter complex and is defined as the stage in binding where the receptor and ligand are now close enough to bind but have not yet done so. RL represents the receptor and its bound ligand. $k_+{}^S$ and $k_-{}^S$ are defined as the transport rate constants for ligand in solution and receptors on the cell surface; a second set of transport rate constants, k_+' and k_-', are used for the formation of the encounter complex. The association and dissociation rate constants k_i and k_{-i} represent the intrinsic rate constants for reaction. It is simplest to think of this three stage binding process as a two stage process of transport to the cell surface and overall reaction with receptors at the surface. This can be

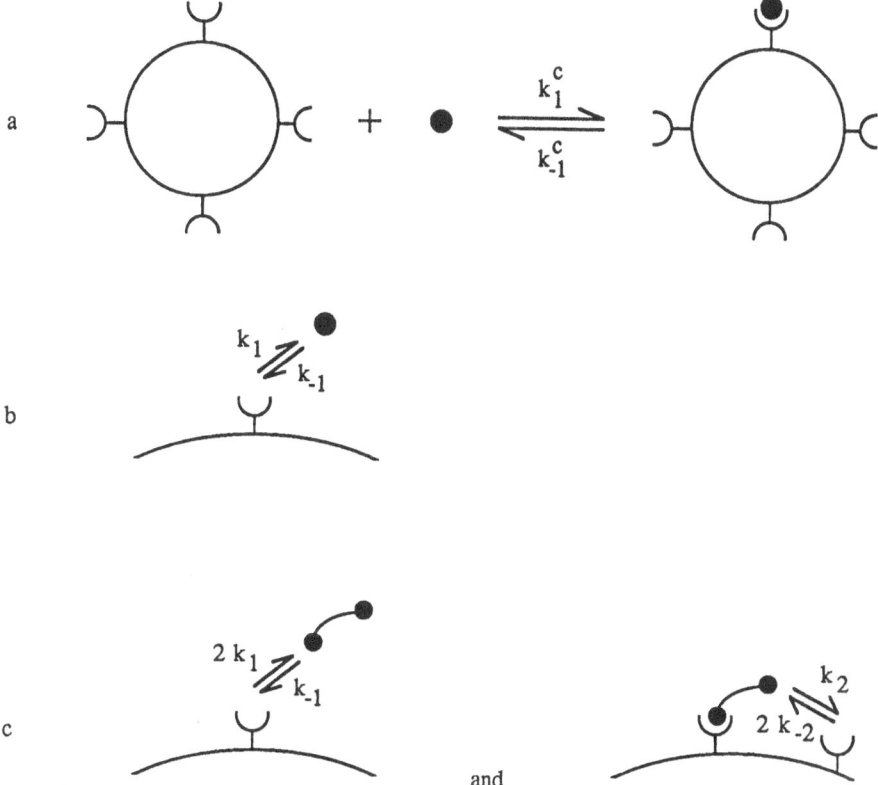

FIGURE A2.1 Binding and crosslinking rate constants. (a) The binding of a monovalent ligand to a cell expressing receptors for that ligand is shown. The measured per cell rate constants are k_1^c for binding and k_{-1}^c for dissociation. (b) The binding and dissociation rate constants for a monovalent ligand and a single receptor are k_1 and k_{-1}, respectively. (c) The binding and crosslinking reactions for a symmetric bivalent ligand and two cell surface receptors are shown. The rate constants k_1 and k_{-2} must be multiplied by a statistical factor of 2 to account for the fact that there are two identical events that could occur in these steps.

shown schematically as

$$R + L \underset{k_-^s}{\overset{k_+^s}{\rightleftharpoons}} R\text{---}L \underset{k_{-1}}{\overset{k_1}{\rightleftharpoons}} RL$$

where R---L is a pseudo-encounter complex of the surface receptors and nearby ligand and k_1 and k_{-1} are the overall per receptor rate constants incorporating the transport and reaction rate constants k_+', k_-', k_i, and k_{-i}. The overall rate constants for ligand binding to the cell are then given by

$$k_1^c = \frac{k_1 k_+^s}{k_1 + k_-^s} = \frac{k_1 k_+^s / k_-^s}{1 + \dfrac{k_1 k_+^s / k_-^s}{k_+^s}} \tag{A2.1}$$

and

$$k_{-1}^c = \frac{k_{-1} k_-^s}{k_1 + k_-^s} = \frac{k_{-1}}{1 + \dfrac{k_1 k_+^s / k_-^s}{k_+^s}} \tag{A2.2}$$

(DeLisi 1980) and are shown also in a rearranged form.

When the binding of ligand to cell surface receptors is reaction-limited, defined by $k_1 \ll k_-^s$ and indicating that the ligand is more likely to leave the cell surface than to bind to a receptor[1], Eq. A2.1 dictates that

$$k_1^c \approx k_1 \left(\frac{k_+^s}{k_-^s}\right) \qquad\qquad k_1 \ll k_-^s \tag{A2.3}$$

(Lauffenburger and DeLisi 1983). Because the binding to the cell surface is assumed reaction-limited, the forward rate constant for the cell, given in Eq. A2.3, is simply equal to

---

[1] The actual binding of ligand to cell surface receptors once the ligand has reached the surface, denoted by k_1 and k_{-1}, may still be either reaction or transport limited (B. Goldstein, personal communication).

Nk_1, where N is the total number of available surface receptors. This substitution can be made in Eqs. A2.1 and A2.2. The forward transport rate constant is also known for the case of transport by pure diffusion to a spherical cell and is given by

$$k_+^s = 4\pi Da \qquad (A2.4)$$

where D is the ligand diffusion coefficient and a is the radius of the cell (DeLisi 1980).

With these substitutions, the overall per cell rate constants become

$$k_1^c = \frac{Nk_1}{1 + \dfrac{Nk_1}{4\pi Da}} \qquad (A2.5)$$

and

$$k_{-1}^c = \frac{k_{-1}}{1 + \dfrac{Nk_1}{4\pi Da}} \qquad (A2.6)$$

(B. Goldstein, personal communication). Thus if the overall per cell rate constants k_1^c and k_{-1}^c are known from experiment, the per receptor rate constants k_1 and k_{-1} can be determined by solving Eqs. A2.5 and A2.6 for these two unknowns. In general, k_1^c is not equal to k_1 and k_{-1}^c is not equal to k_{-1} because of the presence of multiple receptors on a cell and interference between those receptors. DeLisi (1980), Lauffenburger and DeLisi (1983), and Shoup and Szabo (1982) provide further discussion of these rate constants.

Monovalent receptor with bivalent ligand

Using the above formalism, we can also consider the binding of a ligand which is a stable bivalent complex of the monovalent ligand previously discussed. The relevant rate constants for binding and crosslinking can be estimated in the following way. If there is no interference or blocking of the binding sites on the ligand and the diffusion coefficients of the monovalent and bivalent ligands do not differ significantly, the initial binding and dissociation rate constants will be the same as for the monovalent ligand with one slight alteration: the association rate constant k_1 must be multiplied by two, a statistical factor, to account for the fact that there are now two equally accessible binding sites on the ligand.

The rate constants for the crosslinking reaction can be estimated as follows. In analogy with the binding of ligand in solution to cell receptors, crosslinking can also be considered to include transport and reaction steps:

$$
RL \;+\; R \;\;\underset{k_-^{\,m}}{\overset{k_+^{\,m}}{\rightleftharpoons}}\;\; RL\!-\!R \;\;\underset{k_{-i}}{\overset{k_i}{\rightleftharpoons}}\;\; RLR
$$

where $k_+^{\,m}$ and $k_-^{\,m}$ are the forward and reverse transport rate constants in the membrane, RL—R is the encounter complex, and RLR represents a ligand simultaneously bound to two receptors. Again, k_i and k_{-i} represent the intrinsic rate constants for reaction. Lauffenburger and DeLisi (1983) have shown that the transport rate constants when both reacting molecules, in this case a free and a bound receptor, move by pure diffusion in the plane of the membrane are:

$$
k_+^{\,m} = \frac{2\,\pi\,(\,D^m\,(RL) + D^m\,(R)\,)}{\ln\!\left(\dfrac{b}{s}\right) - \dfrac{3}{4}} \qquad\qquad s \ll b \qquad\qquad (A2.7)
$$

and

$$
k_-^{\,m} = \frac{2\,(\,D^m\,(RL) + D^m\,(R)\,)}{s^2\,[\,\ln\!\left(\dfrac{b}{s}\right) - \dfrac{3}{4}\,]} \qquad\qquad s \ll b, \qquad\qquad (A2.8)
$$

where $D^m(i)$ is the two dimension diffusion coefficient of species i in the membrane and s is the radius of the receptor. A reasonable estimate for the half mean distance between receptors, b, can be found from

$$
b \approx \frac{2\,a}{\sqrt{N}} \qquad\qquad (A2.9)
$$

if it is assumed that the N receptors are uniformly distributed over a sphere of radius a. If the case under consideration is the binding of ligand to the cell surface, a is the radius of the cell. The half mean distance between receptors inside the endosome is more difficult to estimate; one possibility is to consider the vesicle as a sphere of radius a and estimate the number of receptors present, N.

We assume in this formulation for k_+^m and k_-^m that the rate limiting step in the formation of the encounter complex RL—R is the translational diffusion of the two reactants. If the reactants are not uniformly reactive over their surfaces and the rotational diffusion necessary to align the binding sites on the two reactants is a significant step in the process, then the above estimates of k_+^m and k_-^m must be modified to account for this fact . In addition, we have neglected here any dependence of the rate constants on the length of the spacer in the ligand molecule that connects the two binding sites, and in some situations this must also be considered (Dembo et al. 1979).

If estimates of the intrinsic rate constants k_i and k_{-i} were available, the transport rate constants k_+^m and k_-^m, together with the intrinsic reaction rate constants, can be used to obtain the overall forward and reverse rate constants for crosslinking according to

$$k_2^* = \frac{k_i \, k_+^m}{k_i + k_-^m} \tag{A2.10}$$

and

$$k_{-2} = \frac{k_{-i} \, k_-^m}{k_i + k_-^m} \tag{A2.11}$$

To estimate the intrinsic rate constants k_i and k_{-i} that are needed, we turn again to the case of monovalent ligand binding to monovalent surface receptors. There is an approximate method for the estimation of k_i and k_{-i} that can be used here, and we next detail this method. The binding of monovalent ligand to monovalent surface receptors can be rewritten as the sum of only two steps, the transport step to the cell surface described earlier and an intrinsic reaction step. Note that we omit the transport step at the cell surface, considering it not to be rate limiting. This can be represented as

$$R \ + \ L \ \underset{k_-^s}{\overset{k_+^s}{\rightleftarrows}} \ R-L \ \underset{k_{-i}}{\overset{k_i}{\rightleftarrows}} \ RL$$

When ligand motion is by pure diffusion, the transport rate constants have been shown theoretically and experimentally to be approximately (DeLisi 1980; Berg and Purcell 1977;

Erickson et al. 1987):

$$k_+^s = 4\pi a D^s \left[\frac{N s}{N s + \pi a} \right]$$ (A2.12)

and

$$k_-^s = \frac{3 D^s}{s^2} \left[\frac{\pi a}{N s + \pi a} \right],$$ (A2.13)

where a is the radius of the cell, D^s is the diffusion coefficient of the ligand in solution, N is the number of free surface receptors, and s is the radius of the receptor. The overall rate constants for ligand binding to the cell are then given by (DeLisi 1980):

$$k_1^c = \frac{k_i k_+^s}{k_i + k_-^s}$$ (A2.14)

$$k_{-1}^c = \frac{k_{-i} k_-^s}{k_i + k_-^s}.$$ (A2.15)

Thus if the overall per cell rate constants k_1^c and k_{-1}^c are known from experiment and k_+^s and k_-^s are estimated from Eqs. A2.12 and A2.13, the intrinsic rate constants for reaction, k_i and k_{-i}, can be determined by solving Eqs. A2.14 and A2.15 for these two unknowns. These intrinsic rate constants can be used together with the calculated estimates for k_+^m and k_-^m to obtain the overall per receptor crosslinking rate constants k_2^* and k_{-2}.

The results of a sample calculation of these rate constants for reasonable values of the many parameters are given in Tables A2.1 (sample calculation at extracellular pH) and A2.2 (endosome pH). Dembo et al. (1982) were able to deduce a value for the forward rate constant for crosslinking in the binding of covalently linked IgE dimers to human basophil Fc receptors by fitting a model to a variety of experiments. Their experimental value agrees in magnitude with our sample calculated value of k_2^*, the forward rate constant for crosslinking.

parameter or rate constant	value
a	5.0×10^{-4} cm
D^m	1.0×10^{-10} cm^2/sec
D^s	1.0×10^{-6} cm^2/sec
N	1.0×10^5 receptors/cell
s	2.0×10^{-7} cm
$k_1{}^c$	1.0×10^9 M^{-1} sec^{-1}
$k_{-1}{}^c$	1.0×10^{-4} sec^{-1}
$k_+{}^s$	3.5×10^{12} M^{-1} sec^{-1}
$k_-{}^s$	5.5×10^6 sec^{-1}
k_i	1.6×10^3 sec^{-1}
k_{-i}	1.0×10^{-4} sec^{-1}
k_1	1.0×10^4 M^{-1} sec^{-1}
k_{-1}	1.0×10^{-4} sec^{-1}
b	3.2×10^{-6} cm
$k_+{}^m$	6.3×10^{-10} cm^2/sec
$k_-{}^m$	5.0×10^3 sec^{-1}
$k_2{}^*$	1.5×10^{-10} cm^2/sec
k_{-2}	7.6×10^{-5} sec^{-1}

TABLE A2.1 Sample calculations of rate constants k_1, k_{-1}, $k_2{}^*$, and k_{-2} at extracellular pH. The values of the first seven parameters and rate constants are assumed to be known from experimental measurements or estimates. D^m(RL) and D^m(R) are assumed equal and are represented by D^m. The rate constants calculated here can be used to calculate the fractions of surface receptors bound and crosslinked as a function of time or at equilibrium. This information can then be used to calculate the initial condition for the single endosome model.

parameter or rate constant	value
k_{-1}^{c}	1.0×10^{-2} sec^{-1}
k_{+}^{s}	3.5×10^{12} M^{-1} sec^{-1}
k_{-}^{s}	5.5×10^{6} sec^{-1}
k_{i}	1.6×10^{3} sec^{-1}
k_{-i}	1.0×10^{-2} sec^{-1}
k_{1}	1.0×10^{4} M^{-1} sec^{-1}
k_{-1}	1.0×10^{-2} sec^{-1}
k_{+}^{m}	6.3×10^{-10} cm^{2}/sec
k_{-}^{m}	5.0×10^{3} sec^{-1}
k_{2}^{*}	1.5×10^{-10} cm^{2}/sec
k_{-2}	7.6×10^{-3} sec^{-1}

TABLE A2.2 Sample calculations of rate constants k_1, k_{-1}, k_2^{*}, and k_{-2} at low (endosome) pH for use in the single endosome model. The parameters a, D^m, D^s, N, and s and the rate constant k_1^c are as listed in Table A1. The overall dissociation rate constant per cell, k_{-1}^c, is assumed to be affected by the low pH and is increased from the value given in Table A2.1 to the value listed here.

SYMBOLS

A_{end}	surface area of the endosome
b	radius of a circular sink
b	radius of an endosome tubule
C	constant
D	general three-dimensional translational diffusion coefficient
D_L	three-dimentional translational diffusion coefficient for ligand molecule
D_n	translational diffusion coefficient in dimension n
D_R	two-dimensional translational diffusion coefficient for receptor molecule
f_b	fraction of receptors bound but not crosslinked
f_c	fraction of receptors crosslinked
f_L	fraction of ligand not recycled from sorting process
$f_L{}^A$	fraction of ligand A not recycled from sorting process
$f_L{}^B$	fraction of ligand B not recycled from sorting process
$f_L{}^{exp}$	experimental fraction of ligand not recycled from sorting process
$f_L{}^{theor}$	theoretical fraction of ligand not recycled from sorting process
f_R	fraction of receptors not recycled from sorting process
$f_R{}^A$	fraction of receptors for ligand A not recycled from sorting process
$f_R{}^B$	fraction of receptors for ligand B not recycled from sorting process
$f_R{}^{exp}$	experimental fraction of receptors not recycled from sorting process
$f_R{}^{theor}$	theoretical fraction of receptors not recycled from sorting process
$F^{(n)}$	fraction of initial number of receptors remaining after n cycles of endocytosis
F_L	dimensionless time for ligand to move from endosome vesicle to tubule
F_R	dimensionless time for receptor to move from endosome vesicle to tubule
G	Green's function
j	average value of derivative given in Eq. 4.17
k	synthesis rate of new receptors
k_1	rate constant for binding of receptor and ligand at endosome pH
k_{-1}	rate constant for dissociation of receptor/ligand complex at endosome pH
$k_1{}^N$	rate constant for binding of receptor and ligand at extracellular pH
$k_{-1}{}^N$	rate constant for dissociation of receptor/ligand complex at extracellular pH
k_2	rate constant for crosslinking at endosome pH

k_{-2}	rate constant for release of crosslink at endosome pH
k_2^N	rate constant for crosslinking at extracellular pH
k_{-2}^N	rate constant for release of crosslink at extracellular pH
k_2^*	rate constant for crosslinking at endosome pH with units area/time
k_2^{*N}	rate constant for crosslinking at extracellular pH with units area/time
k_a	rate constant for free receptor movement from vesicle to tubule
k_b	rate constant for bound (uncrosslinked) receptor movement from vesicle to tubule
k_c	rate constant for crosslinked receptor movement from vesicle to tubule
k_D	rate constant for lysosomal degradation and release of products
k_D^L	rate constant for lysosomal degradation of ligand and release of products
k_I	rate constant for internalization
k_I^L	rate constant for internalization of ligand
k_L	rate constant for movement of material from endosome to lysosome
k_R	rate constant for movement of material from endosome to cell surface
k_S	rate constant for synthesis of receptors
K_1^N	equilibrium constant for binding of receptor and ligand at extracellular pH
K_2^N	equilibrium constant for crosslinking at extracellular pH
K_D	dissociation equilibrium constant
L_0	concentration of ligand in medium
L_1	number of bound (uncrosslinked) ligands in vesicle
$\overline{L_1}$	dimensionless number of bound (uncrosslinked) ligands in vesicle
$L_1^{/}$	number of bound (uncrosslinked) ligands in tubule
$\overline{L_1}^{/}$	dimensionless number of bound (uncrosslinked) ligands in tubule
L_2	number of ligand molecules crosslinking two receptors in vesicle
$\overline{L_2}$	dimensionless number of ligand molecules crosslinking two receptors in vesicle
$L_2^{/}$	number of ligand molecules crosslinking two receptors in tubule
$\overline{L_2}^{/}$	dimensionless number of ligand molecules crosslinking two receptors in tubule
L_d	dimensionless number of degraded ligand molecules
L_f	concentration of free ligand in vesicle
$L_f^{/}$	concentration of free ligand in tubule
L_i	dimensionless number of ligand molecules in endosome
L_l	dimensionless number of ligand molecules in lysosome
L_s	dimensionless number of bound ligand molecules on the cell surface
L_T	total number of ligand molecules in endosome
n	number of endocytotic cycles
N_A	Avogadro's number
r	radial coordinate

r'	radial coordinate
r	observation point in Green's function
r'	source point in Green's function
R	radius of the vesicular part of an endosome
R_0	total receptors at time zero
R_1	number of bound receptors in vesicle
R_1	ratio of receptors recycled to ligand recycled
$R_1{}^/$	number of bound receptors in tubule
R_2	ratio of receptor degraded to ligand degraded
R_d	dimensionless number of degraded receptors
$R_d{}^0$	initial dimensionless number of receptors that have been degraded
R_f	number of free receptors in vesicle
$\overline{R_f}$	dimensionless number of free receptors in vesicle
$R_f{}^/$	number of free receptors in tubule
$\overline{R_f{}^/}$	dimensionless number of free receptors in tubule
R_i	dimensionless number of receptors in endosome
$R_i{}^0$	initial dimensionless number of receptors in endosome
R_l	dimensionless number of receptors in lysosome
$R_l{}^0$	initial dimensionless number of receptors in lysosome
R_s	dimensionless number of bound surface receptors
$R_s{}^0$	initial dimensionless number of bound surface receptors
R_T	total number of receptors in endosome
S_0	density of surface receptors
t	time
t_0	time zero
v	membrane velocity
$\overline{v}$	average membrane velocity
V	total volume of endocytic vesicles internalized per unit time
V_T	tubule volume
V_V	vesicle volume
W	mean capture time
W_L	mean capture time for ligand molecule
$\overline{W_L}$	volume averaged mean capture time for ligand molecule
$W_L{}^{(surf)}$	surface averaged mean capture time for ligand molecule
$\overline{W_L}{}^{\Gamma P}$	averaged mean capture time for diffusion to a hole on an infinite plane
$\overline{W_L}{}^H$	averaged mean capture time for diffusion within a hemisphere to a smaller concentric hemisphere

W_R	mean capture time for receptor molecule
$\overline{W_R}$	surface averaged mean capture time for receptor molecule
x	Cartesian coordinate
y	Cartesian coordinate

Greek letters

α	fraction of the total vesicle area removed per unit time by membrane current
α_1	dimensionless integral defined by Eq. 4.27
α_2	dimensionless integral defined by Eq. 4.28
β	dimensionless current
β	dimensionless parameter defined by Eq. 6.42
β_1	dimensionless integral defined by Eq. 4.22
β_1'	dimensionless integral defined by Eq. 4.29
β_2	dimensionless integral defined by Eq. 4.23
β_2'	dimensionless integral defined by Eq. 4.30
γ	angle between (θ,ϕ) and (θ',ϕ')
$\gamma1$	dimensionless transport parameter describing the movement of bound receptors
$\gamma2$	dimensionless transport parameter describing the movement of crosslinked receptors
δ	dimensionless parameter defined by Eq. 6.43
δ	displacement
Δ_n^2	Laplacian operator in dimension n
Δt	time increment
ε	ratio of dissociation rate constant to rate constant for movement of free receptors
η	dimensionless equilibrium binding constant
θ	angular coordinate
θ	endocytosis cycle time
θ'	angular coordinate
$\theta*$	angular position
θ_c	critical angle defining the size of the tubule opening
κ	partition coefficient
λ	diameter of ligand molecule/diameter of tubule
ξ	dimensionless radial position
ξ'	dimensionless radial position

ρ	radius of hemisphere placed around a sink
ρ	ratio of total ligand molecules to total receptor molecules in endosome
σ	dimensionless rate of breakup of crosslinks
τ	dimensionless time
τ^*	dimensionless sorting time
ϕ	angular coordinate
ϕ'	angular coordinate
ϕ^*	angular position
Φ	state transition matrix
χ	dimensionless rate of crosslinking

REFERENCES

Abercrombie, M., J.E.M. Heaysman, and S.M. Pegrum. 1970. The locomotion of fibroblasts in culture. III. Movements of particles on the dorsal surface of the leading lamella. *Exp. Cell Res.* 62: 389-398.

Adam, G., and M. Delbrück. 1968. Reduction of dimensionality in biological diffusion processes. In *Structural Chemistry and Molecular Biology.* A. Rich and N. Davidson, editors. W.H. Freeman and Co., San Francisco. 198-215.

Alberts, B., D. Bray, J. Lewis, M. Raff, K. Roberts, and J.D. Watson. 1983. *Molecular Biology of the Cell.* Garland Publishing, Inc. New York.

Anderson, R. G. W., M. S. Brown, U. Beisiegel, and J. L. Goldstein. 1982. Surface distribution and recycling of the low density lipoprotein receptor as visualized with antireceptor antibodies. *J. Cell Biol.* 93: 523-531.

Ashwell, G. and J. Harford. 1982. Carbohydrate-specific receptors of the liver. *Ann. Rev. Biochem.* 51: 531-554.

Baenziger, J. U. and D. Fiete. 1986. Separation of two populations of endocytic vesicles involved in receptor-ligand sorting in rat hepatocytes. *J. Biol. Chem.* 261: 7445-7454.

Basu, M., K. Frick, A. Sen-Majumdar, C.D. Scher, and M. Das. 1985. EGF receptor-associated DNA-nicking activity is due to a M_r-100,000 dissociable protein. *Nature* 316: 640-641.

Basu, S.K., J. L. Goldstein, R.G.W. Anderson, and M.S. Brown. 1981. Monensin interrupts the recycling of low density lipoprotein receptors in human fibroblasts. *Cell* 24: 493-502.

Basu, S.K., J. L. Goldstein, and M.S. Brown. 1978. Characterization of the low density lipoprotein receptor in membranes prepared from human fibroblasts. *J. Biol. Chem.* 253: 3852-3856.

Berg, H. C., and E. M. Purcell. 1977. Physics of chemoreception. *Biophys. J.* 20: 193-219.

Bierer, B.E., S.H. Herrmann, C.S. Brown, S.J. Burakoff, and D.E. Golan. 1987. Lateral mobility of class I histocompatibility antigens in B lymphoblastoid cell membranes: Modulation by cross-linking and effect of cell density. *J. Cell Biol.* 105: 1147-1152.

Braell, W. A. 1987. Fusion between endocytic vesicles in a cell-free system. *Proc. Natl. Acad. Sci. USA* 84: 1137-1141.

Bretscher, M.S. 1976. Directed lipid flow in cell membranes. *Nature (Lond.)* 260: 21-23.

Bretscher, M.S. 1984. Endocytosis: Relation to capping and cell locomotion. *Science* 244: 681-686.

Bretscher, M. S., and J. N. Thomson. 1985. The morphology of endosomes in giant HeLa cells. *Eur. J. Cell Biol.* 37: 78-80.

Bridges, K., J. Harford, G. Ashwell, and R.D. Klausner. 1982. Fate of receptor and ligand during endocytosis of asialglycoproteins by isolated hapatocytes. *Proc. Natl. Acad. Sci. USA* 79: 350-354.

Brown, M.S., R.G.W. Anderson, and J.L. Goldstein. 1983. Recycling receptors: The round-trip itinerary of migrant membrane proteins. *Cell* 32: 663-667.

Brown, M.S., and J.L. Goldstein. 1986. A receptor-mediated pathway for cholesterol homeostasis. *Science* 232: 34-47.

Brown, W.J., J. Goodhouse, and M.G. Farquhar. 1986. Mannose-6-phosphate receptors for lysosomal enzymes cycle between the Golgi complex and endosomes. *J. Cell Biol.* 103: 1235-1247.

Brunn, P.O. 1981. Absorption by bacterial cells: Interaction between receptor sites and the effect of fluid motion. *J. Biomech. Eng.* 103: 32-37.

Ciechanover, A., A. L. Schwartz, A. Dautry-Varsat, and H. F. Lodish. 1983. Kinetics of internalization and recycling of transferrin and the transferrin receptor in a human hepatoma cell line: Effect of lysosomotropic agents. *J. Biol. Chem.* 258: 9681-9689.

Ciechanover, A., A. L. Schwartz, and H. L. Lodish. 1985. Sorting and recycling of cell surface receptors and endocytosed ligands: The asialoglycoprotein and transferrin receptors. In *Mechanisms of Receptor Regulation.* G. Poste and S. T. Crooke, editors. Plenum Press, New York. 225-253.

Corin, R.E., and D.B. Donner. 1982. Insulin receptors convert to a higher affinity state subsequent to hormone binding: A two-state model for the insulin receptor. *J. Biol. Chem.* 257: 104-110.

Cuatrecasas, P. 1982. Epidermal growth factor: Uptake and fate. In *Membrane Recycling.* (Ciba Foundation Symposium 92) Pitman Books, Ltd., London. 96-108.

Daukas, G., D.A. Lauffenburger, and S. Zigmond. 1983. Reversible pinocytosis in polymorphonuclear leukocytes. *J. Cell Biol.* 96: 1642-1650.

Davey, J., S.M. Hurtley, and G. Warren. 1985. Reconstitution of an endocytic fusion event in a cell-free system. *Cell* 43: 643-652.

Davis, C.G., J.L. Goldstein, T.C. Südhof, R.G.W. Anderson, D.W. Russell, and M.S. Brown. 1987. Acid-dependent ligand dissociation and recycling of LDL receptor mediated by growth factor homology region. *Nature* 326: 760-765.

DeLisi, C. 1980. The biophysics of ligand-receptor interactions. *Quart. Revs. of Biophys.* 13: 201-230.

DeLisi, C., and F.W. Wiegel. 1981. Effect of nonspecific forces and finite receptor number on rate constants of ligand-cell bound-receptor interactions. *Proc. Natl. Acad. Sci. USA* 78: 5569-5572.

Dembo, M., B. Goldstein, A. K. Sobotka, and L. M. Lichtenstein. 1979. Histamine release due to bivalent penicilloyl haptens: The relation of activation and desensitization of basophils to dynamic aspects of ligand binding to cell surface antibody. *J. Immunol.* 122: 518-528.

Dembo, M., A. Kagey-Sobotka, L.M. Lichtenstein, and B. Goldstein. 1982. Kinetic analysis of histamine release due to covalently linked IgE dimers. *Molec. Immunol.* 19: 421-434.

DiPaola, M. and F.R. Maxfield. 1984. Conformational changes in the receptors for epidermal growth factor and asialoglycoproteins induced by the mildly acidic pH found in endocytic vesicles. *J. Biol. Chem.* 259: 9163-9171.

Dunn, W. A., and A. L. Hubbard. 1984. Receptor-mediated endocytosis of epidermal growth factor by hepatocytes in the perfused rat liver: Ligand and receptor dynamics. *J. Cell Biol.* 98: 2148-2159.

Earp, H.S., K.S. Austin, J. Blaisdell, R.A. Rubin, K.G. Nelson, L.W. Lee, and J.W. Grisham. 1986. Epidermal growth factor (EGF) stimulates EGF receptor synthesis. *J. Biol. Chem.* 261: 4777-4780.

Edidin, M., Y. Zagyansky, and T.J. Lardner. 1976. Measurement of membrane protein lateral diffusion in single cells. *Science* 191: 466-468.

Einstein, A. 1926. *Investigations on the Theory of the Brownian Movement.* Dover Publications, Inc. Dover Publications, Inc.

Erickson, J., B. Goldstein, D. Holowka, and B. Baird. 1987. The effect of receptor density on the forward rate constant for binding of ligands to cell surface receptors. *Biophys. J.* 52: 657-662.

Evans, E.A., and R. Skalak. 1980. *Mechanics and Thermodynamics of Biomembranes.* CRC Press, Inc., Boca Raton, FL.

Evans, W.H. and N. Flint. 1985. Subfractionation of hepatic endosomes in Nycodenz gradients and by free-flow electrophoresis: Separation of ligand-transporting and receptor-enriched membranes. *Biochem. J.* 232: 25-32.

Ezaki, O., M. Kasuga, Y. Akanuma, K. Takata, H. Hirano, Y. Fujita-Yamaguchi, and M. Kasahara. 1986. Recycling of the glucose transporter, the insulin receptor, and insulin in rat adipocytes: Effect of acidtropic agants. *J. Biol. Chem.* 261: 3295-3305.

Furuichi, K., J. Rivera, L.M. Buonocore, and C. Isersky. 1986. Recycling of receptor-bound IgE by rat basophilic leukemia cells. *J. Immunol.* 136: 1015-1022.

Furuichi, K., J. Rivera, T. Triche, and C. Isersky. 1985. The fate of IgE bound to rat basophilic leukemia cells. IV. Functional association between the receptors for IgE. *J. Immunol.* 134: 1766-1773.

Galloway, C.J., G.E. Dean, M. Marsh, G. Rudnick, and I. Mellman. 1983. Acidification of macrophage and fibroblast endocytic vesicles *in vitro. Proc. Natl. Acad. Sci. USA* 80: 3334-3338.

Gear, C.W. 1971. *Numerical Initial Value Problems in Ordinary Differential Equations.* Prentice-Hall, Englewood Cliffs, NJ.

Geuze, H.J., J.W. Slot, and A.L. Schwartz. 1987. Membranes of sorting organelles display lateral heterogeneity in receptor distribution. *J. Cell Biol.* 104: 1715-1723.

Geuze, H. J., J. W. Slot, G. J. A. M. Strous, H. F. Lodish, and A. L. Schwartz. 1983. Intracellular site of asialoglycoprotein receptor-ligand uncoupling: Double-label immunoelectron microscopy during receptor-mediated endocytosis. *Cell* 32: 277-287.

Geuze, H. J., J. W. Slot, G. J. A. M. Strous, J. Peppard, K. von Figura, A. Hasilik, and A.L. Schwartz. 1984. Intracellular receptor sorting during endocytosis: Comparative immunoelectron microscopy of multiple receptors in rat liver. *Cell* 37: 195-204.

Gex-Fabry, M., and C. DeLisi. 1984. Model for kinetic and steady state analysis of receptor mediated endocytosis. *Math. Biosci.* 72: 245-261.

Giugni, T.D., D. L. Braslau, and H.T. Haigler. 1987. Electric field-induced redistribution and postfield relaxation of epidermal growth factor receptors on A431 cells. *J. Cell Biol.* 104: 1291-1297.

Goldstein, B., R. Griego, and C. Wofsy. 1984. Diffusion-limited forward rate constants in two dimensions: Application to the trapping of cell surface receptors by coated pits. *Biophys. J.* 46: 573-586.

Goldstein, J.L., R.G.W. Anderson, and M.S. Brown. 1982. Receptor-mediated endocytosis and the cellular uptake of low density lipoprotein. In *Membrane Recycling.* (Ciba Foundation Symposium 92) Pitman Books, Ltd., London. 77-95.

Gonzalez-Noriega, A., J.H. Grubb, V. Talkad, and W.S. Sly. 1980. Chloroquine inhibits lysosomal enzyme pinocytosis and enhances lysosomal enzyme secretion by impairing receptor recycling. *J. Cell Biol.* 85: 839-852.

Greenspan, P., and R.W. St. Clair. 1984. Retroendocytosis of low density lipoprotein: Effect of lysosomal inhibitors on the release of undegraded ^{125}I-low density lipoprotein of altered composition from skin fibroblasts in culture. *J. Biol. Chem.* 259: 1703-1713.

Gruenberg, J.E., and K.E. Howell. 1986. Reconstitution of vesicle fusions occurring in endocytosis with a cell-free system. *EMBO J.* 5: 3091-3101.

Harding, C., M. A. Levy, and P. Stahl. 1985. Morphological analysis of ligand uptake and processing: The role of multivesicular endosomes and CURL in receptor-ligand processing. *Eur. J. Cell Biol.* 36: 230-238.

Harford, J., A.W. Wolkoff, G. Ashwell, and R.D. Klausner. 1983. Monensin inhibits dissociation of asialoglycoproteins from their receptor. *J. Cell Biol.* 96: 1824-1828.

Hatae, T., M. Fujita, H. Sagara, and K. Okuyama. 1986. Formation of apical tubules from large endocytic vacuoles in kidney proximal tubule cells during absorption of horseradish peroxidase. *Cell and Tissue Res.* 246: 271-278.

Helenius, A., I. Mellman, D. Wall, and A. Hubbard. 1983. Endosomes. *Trends Biochem. Sci.* 8: 245-250.

Hill, T.L. 1975. Effect of rotation on the diffusion-controlled rate of ligand-protein association. *Proc. Natl. Acad. Sci. USA.* 72: 4918-4922.

Hillman, G.M., and J. Schlessinger. 1982. Lateral diffusion of epidermal growth factor complexed to its surface receptors does not account for the thermal sensitivity of patch formation and endocytosis. *Biochemistry* 21: 1667-1672.

Hopkins, C. R., and I. S. Trowbridge. 1983. Internalization and processing of transferrin and the transferrin receptor in human carcinoma A431 cells. *J. Cell Biol.* 97: 508-521.

Hudgin, R.L., W.E. Pricer, Jr., G. Ashwell, R.J. Stockert, and A.G. Morell. 1974. The isolation and properties of a rabbit liver binding protein specific for asialoglycoproteins. *J. Biol. Chem.* 249: 5536-5543.

Huecksteadt, T., J.M. Olefsky, D. Brandenberg, and K.A. Heidenreich. 1986. Recycling of photoaffinity-labeled insulin receptors in rat adipocytes: Dissociation of insulin-receptor complexes is not required for receptor recycling. *J. Biol. Chem.* 262: 8655-8659.

Isersky, C., J. Rivera, S. Mims, and T. J. Triche. 1979. The fate of IgE bound to rat basophilic leukemia cells. *J. Immunol.* 122: 1926-1936.

Kelly, R.B. 1985. Pathways of protein secretion in eukaryotes. *Science* 230: 25-32.

Kielian, M.C., M. Marsh, and A. Helenius. 1986. Kinetics of endosome acidification detected by mutant and wild-type Semliki Forest virus. *The EMBO J.* 5: 3103-3109.

Klausner, R.D., G. Ashwell, J. van Renswoude, J.B. Harford, and K.R. Bridges. 1983a. Binding of apotransferrin to K562 cells: Explanation of the transferrin cycle. *Proc. Natl. Acad. Sci. USA* 80: 2263-2266.

Klausner, R.D., J.Van Renswoude, G. Ashwell, C. Kempf, A.N. Schechter, A. Dean, and K.R. Bridges. 1983b. Receptor-mediated endocytosis of transferrin in K562 cells. *J. Biol. Chem.* 258: 4715-4724.

Koshlyakov, N.S., M.M. Smirnov, and E.B. Gliner. 1964. *Differential Equations of Mathematical Physics.* North-Holland Publishing Co., Amsterdam. 350.

Lauffenburger, D. and C. DeLisi. 1983. Cell surface receptors: Physical chemistry and cellular regulation. *International Rev. of Cytology* 84: 269-302.

Lauffenburger, D., J. Linderman, and L. Berkowitz. 1987. Analysis of mammalian cell growth factor receptor dynamics. *Ann. NY Acad. Sci.* 506: 147-162.

Macken, C.A., and A. S. Perelson. 1985. Branching processes applied to cell surface aggregation phenomena. *Lecture Notes in Biomathematics* 58. Springer-Verlag, Berlin.

Marsh, M. 1984. The entry of enveloped viruses into cells by endocytosis. *Biochem. J.* 218: 1-10.

Marsh, M., G. Griffiths, G. E. Dean, I. Mellman, and A. Helenius. 1986. Three-dimensional structure of endosomes in BHK-21 cells. *Proc. Natl. Acad. Sci. USA.* 83: 2899-2903.

Marsh, M., S. Schmid, H. Kern, E. Harms, P. Male, I. Mellman, and A. Helenius. 1987. Rapid analytical and preparative isolation of functional endosomes by free flow electrophoresis. *J. Cell Biol.* 104: 875-886.

Marshall, S. 1985a. Degradative processing of internalized insulin in isolated adipocytes. *J. Biol. Chem.* 260: 13517-13523.

Marshall, S. 1983. Kinetics of insulin receptor biosynthesis and membrane insertion: Relationship to cellular function. *Diabetes* 32: 319-325.

Marshall, S. 1985b. Kinetics of insulin receptor internalization and recycling in adipocytes: Shunting of receptors to a degradative pathway by inhibitors of recycling. *J. Biol. Chem.* 260: 4136-4144.

Mellman, I., R. Fuchs, and A. Helenius. 1986. Acidification of the endocytic and exocytic pathways. *Ann. Rev. Biochem.* 55: 663-700.

Mellman, I., and H. Plutner. 1984. Internalization and degradation of macrophage Fc receptors bound to polyvalent immune complexes. *J. Cell Biol.* 98: 1170-1177.

Mellman, I.S., H. Plutner, R.M. Steinman, J.C. Unkeless, and Z.A. Cohn. 1983. Internalization and degradation of macrophage Fc receptors during receptor-mediated phagocytosis. *J. Cell Biol.* 96: 887-895.

Mellman, I., H. Plutner, and P. Ukkonen. 1984. Internalization and rapid recycling of macrophage Fc receptors tagged with monovalent antireceptor antibody: Possible role of a prelysosomal compartment. *J. Cell Biol.* 98: 1163-1169.

Mellman, I.S. and J. C. Unkeless. 1980. Purification of a functional mouse Fc receptor through the use of a monoclonal antibody. *J. Exp. Med.* 152: 1048-1069.

Mellman, I.S., R.M. Steinman, J.C. Unkeless, and Z.A. Cohn. 1980. Selective iodination and polypeptide composition of pinocytic vesicles. *J. Cell Biol.* 86: 7112-722.

Menon, A. K., D. Holowka, W. W. Webb, and B. Baird. 1986a. Clustering, mobility, and triggering activity of small oligomers of immunoglobulin E on rat basophilic leukemia cells. *J. Cell Biol.* 102: 534-540.

Menon, A. K., D. Holowka, W. W. Webb, and B. Baird. 1986b. Crosslinking of receptor-bound IgE to aggregated larger than dimers leads to rapid immobilization. *J. Cell Biol.* 102: 541-550.

Mueller, S. C. and A. L. Hubbard. 1986. Receptor-mediated endocytosis of asialoglycoproteins by rat hepatocytes: Receptor-positive and receptor-negative endosomes. *J. Cell Biol.* 102: 932-942.

Murphy, R. F. 1985. Analysis and isolation of endocytic vesicles by flow cytometry and sorting: Demonstration of three kinetically distinct compartments involved in fluid-phase endocytosis. *Proc. Natl. Acad. Sci. USA* 82: 8523-8526.

Murphy, R.F., S. Powers, and C.R. Cantor. 1984. Endosome pH measured in single cells by dual fluorescence flow cytometry: Rapid acidification of insulin to pH 6. *J. Cell Biol.* 98: 1757-1762.

Murphy, R.F. and M. Roederer. 1986. Flow cytometric analysis of endocytic pathways. In *Applications of Flourescence in the Biomedical Sciences.* Alan R. Liss, Inc. 545-566.

Olsnes, S., and A. Pihl. 1982. Chimeric toxins. *Pharmac. Ther.* 15: 355-381.

Pappenheimer, J. R. 1953. Passage of molecules through capillary walls. *Physiol. Rev.* 33: 387-423.

Perelson, A., and C. DeLisi. 1980. Receptor clustering on a cell surface. I. Theory of receptor cross-linking by ligands bearing two chemically identical functional groups. *Math. Biosci.* 48: 71-110.

Pfeffer, L.M., N. Stebbing, and D.B. Donner. 1987. Cytoskeletal association of human α-interferon-receptor complexes in interferon-sensitive and -resistant lymphoblastoid cells. *Proc. Natl. Acad. Sci. USA* 84: 3249-3253.

Pfeffer, S.R., and J.E. Rothman. 1987. Biosynthetic protein transport and sorting by the endoplasmic reticulum and Golgi. *Ann. Rev. Biochem.* 56: 829-852.

Powell, M.J.D. 1977. A fast algorithm for nonlinearly constrained optimization calculations. Dundee Conference on Numerical Analysis.

Ramkrishna, D., and N.R. Amundson. 1985. *Linear Operator Methods in Chemical Engineering with Applications to Transport and Chemical Reaction Systems*. Prentice-Hall, Inc., Englewood Cliffs, New Jersey. p. 209.

Robertson, D., D. Holowka, and B. Baird. 1986. Cross-linking of immunoglobulin E-receptor complexes induces their interaction with the cytoskeleton of rat basophilic leukemia cells. *J. Immunol.* 136: 44565-4572.

Robinson, M.S. 1987. 100 kD coated vesicle proteins: Molecular heterogeneity and intracellular distribution studied with monoclonal antibodies. *J. Cell Biol.* 104: 887-895.

Roederer, M., Bowser, R., and Murphy, R. 1987. Kinetics and temperature dependence of exposure of endocytosed material to proteolytic enzymes and low pH: Evidence for a maturation model for the formation of lysosomes. *J. Cell. Physiol.* 131: 200-209.

Schaudies, R.P., R.A. Harper, and C.R. Savage, Jr. 1985. ^{125}I-EGF binding to responsive and nonresponsive cells in culture: Loss of cell-associated radioactivity relates to growth induction. *J. Cell. Physiol.* 124: 493-498.

Schiff, J. M., M. M. Fisher, A.L. Jones, and B. J. Underdown. 1984. Human IgA as a heterovalent ligand: Switching from the asialoglycoprotein receptor to secretory component during transport across the rat hepatocyte. *J. Cell Biol.* 102: 920-931.

Schlessinger, J., Y. Schechter, P. Cuatrecasas, M.C. Willingham, and I. Pastan. 1978. Quantitative determination of the lateral diffusion coefficients of the hormone-receptor complexes of insulin and epidermal growth factor on the plasma membrane of cultured fibroblasts. *Proc. Natl. Acad. Sci. USA* 75: 5353-5357.

Schwartz, A.L., A. Ciechanover, S. Merritt, and A. Turkewitz. 1986. Antibody-induced receptor loss: Different fates for asialoglycoproteins and the asialoglycoprotein receptor in Hep G2 cells. *J. Biol. Chem.* 261: 15225-15232.

Schwartz, A.L., A. Bolognesi, and S.E. Fridovich. 1984. Recycling of the asialoglycoprotein receptor and the effect of lysosomotropic amines in hepatoma cells. *J. Cell Biol.* 98: 732-738.

Schwartz, A. L., S. E. Fridovich, and H. F. Lodish. 1982. Kinetics of internalization and recycling of the asialoglycoprotein receptor in a hepatoma cell line. *J. Biol. Chem.* 257: 4230-4237.

Shoup, D., G. Lipari, A. Szabo. 1981. Diffusion-controlled biomolecular reaction rates: The effect of rotational diffusion and orientation constraints. *Biophys. J.* 36: 697-714.

Shoup, D., and A. Szabo. 1982. Role of diffusion in ligand binding to macromolecules and cell-bound receptors. *Biophys. J.* 40: 33-39.

Stahl, P., P.H. Schlesinger, E. Sigardson, J.S. Rodman, and Y.C. Lee. 1980. Receptor-mediated pinocytosis of mannose glycoconjugates by macrophages: Characterization and evidence for receptor recycling. *Cell* 19: 207-215.

Stahl, P., and A.L. Schwartz. 1986. Receptor-mediated endocytosis. *J. Clin. Invest.* 77: 657-662.

Steinman, R.M., S.E. Brodie, and Z.A. Cohn. 1976. Membrane flow during pinocytosis: A stereologic analysis. *J. Cell Biol.* 68: 665-687.

Steinman, R. M., I. S. Mellman, W. A. Muller, and Z. A. Cohn. 1983. Endocytosis and the recycling of plasma membrane. *J. Cell Biol.* 96: 1-27.

Stoscheck, C.M., and G. Carpenter. 1984. Down regulation of epidermal growth factor receptors: Direct demonstration of receptor degradation in human fibroblasts. *J. Cell Biol.* 98: 1048-1053.

Szabo, A., K. Schulten, and Z. Schulten. 1980. First passage time approach to diffusion controlled reactions. *J. Chem. Phys.* 72: 4350-4357.

Tanford, C. 1961. Physical Chemistry of Macromolecules. John Wiley & Sons., Inc. New York.

Tietze, C., P. Schlesinger, and P. Stahl. 1982. Mannose-specific endocytosis receptor of alveolar macrophages: Demonstration of two functionally distinct intracellular pools of receptor and their roles in receptor recycling. *J. Cell Biol.* 92: 417-424.

Telfer, W.H., E. Huebner, and D.S. Smith. 1982. The cell biology of vitellogenic follicles in *Hyalophora* and *Rhodnius*. In *Insect Ultrastructure*, Vol. 1. King and Akai, eds. pp. 118-149.

Townsend, R. R., D. A. Wall, A. L. Hubbard, and Y. C. Lee. 1984. Rapid release of galactose-terminated ligands after endocytosis by hepatic parenchymal cells: Evidence for a role of carbohydrate structure in the release of internalized ligand from receptor. *Proc. Natl. Acad. Sci. USA.* 81: 466-470.

Tycko, B. and F.R. Maxfield. 1982. Rapid acidification of endocytic vesicles containing α_2-macroglobulin. *Cell* 28: 643-651.

Tycko, B., C. H. Keith, and F. R. Maxfield. 1983. Rapid acidification of endocytic vesicles containing asialoglycoprotein in cells of a human hepatoma line. *J. Cell Biol.* 97: 1762-1776.

Ukkonen, P., V. Lewis, M. Marsh, A. Helenius, and I. Mellman. 1986. Transport of macrophage Fc receptors and Fc receptor-bound ligands to lysosomes. *J. Exp. Med.* 163: 952-971.

van Leuven, F., J. Cassiman, and H. van den Berghe. 1980. Primary amines inhibit recycling of α_2M receptors in fibroblasts. *Cell* 20: 37-43.

van Renswoude, J., K.R. Bridges, J.B. Harford, and R.D. Klausner. 1982. Receptor-mediated endocytosis of transferrin and the uptake of Fe in K562 cells: Identification of a nonlysosomal acidic compartment. *Proc. Natl. Acad. Sci. USA* 79: 6186-6190.

Wall, D. A., G. Wilson, and A. L. Hubbard. 1980. The galactose-specific recognition system of mammalian liver: The route of ligand internalization in rat hepatocytes. *Cell* 21: 79-93.

Watts, C. 1985. Rapid endocytosis of the transferrin receptor in the absence of bound transferrin. *J. Cell Biol.* 100: 633-637.

Weigel, P.H., and J.A. Oka. 1981. Temperature dependence of endocytosis mediated by the asialoglycoprotein receptor in isolated rat hepatocytes. *J. Biol. Chem.* 256: 2615-1617.

Weinberger, H.F. 1965. *A First Course in Partial Differential Equations*. John Wiley & Sons, New York. 195.

Weissman, A. M., R. D. Klausner, K. Rao, and J. B. Harford. 1986. Exposure of K562 cells to anti-receptor monoclonal antibody OKT9 results in rapid redistribution and enhanced degradation of the transferrin receptor. *J. Cell Biol.* 102: 951-958.

Westcott, K.R., R.P. Searles, and L.H. Rome. 1987. Evidence for ligand- and pH-dependent conformational changes in liposome-associated mannose 6-phosphate receptor. *J. Biol. Chem.* 262: 6101-6107.

Wiegant, F. A. C., F. J. Blok, L. H. K. Defize, W. A. M. Linnemans, A. J. Verkley, and J. Boonstra. 1986. Epidermal growth factor receptors associated to cytoskeletal elements of epidermoid carcinoma (A431) cells. *J. Cell Biol.* 103: 87-94.

Wiegel, F. W. 1984. Diffusion of proteins in membranes. In *Cell Surface Dynamics: Concepts and Models*. A.S. Perelson, C. DeLisi, and F.W. Wiegel, editors. Marcel Dekker, Inc. New York. 135-150.

Wiegel, F.W. 1980. Fluid Flow Through Porous Macromolecular Systems. I. Lecture Notes in Physics. Vol 121. Springer-Verlag. New York. 56-60.

Wiegel, F.W. 1979. Hydrodynamics of a permeable patch in the fluid membrane. *J. Theor. Biol.* 77: 189-193.

Wileman, T., C. Harding, and P. Stahl. 1985a. Receptor-mediated endocytosis. *Biochem. J* 232: 1-14.

Wileman, T., R. Boshans, and P. Stahl. 1985b. Uptake and transport of mannosylated ligands by alveolar macrophages: Studies on ATP-dependent receptor-ligand dissociation. *J. Biol. Chem.* 260: 7387-7393.

Wolkoff, A. W., R. D. Klausner, G. Ashwell, and J. Harford. 1984. Intracellular segregation of asialoglycoproteins and their receptor: A prelysosomal event subsequent to dissociation of the ligand-receptor complex. *J. Cell Biol.* 98: 375-381.

Yamashiro, D. J., and F. R. Maxfield. 1984. Acidification of endocytic compartments and the intracellular pathways of ligands and receptors. *J. Cell. Biochem.* 26: 231-246.

Zigmond, S. H., S. J. Sullivan, and D. A. Lauffenburger. 1982. Kinetic analysis of chemotactic peptide receptor modulation. *J. Cell Biol.* 92: 34-43.

INDEX

Your source for advances in theoretical biology and biomathematics

Journal of Mathematical Biology

ISSN 0303-6812   Title No. 285

Editorial Board: K. P. Hadeler, Tübingen; S. A. Levin, Ithaca (Managing Editors); H. T. Banks, Providence; J. D. Cowan, Chicago; J. Gani, Santa Barbara; F. C. Hoppenstedt, East Lansing; D. Ludwig, Vancouver; J. D. Murray, Oxford; T. Nagylaki, Chicago; L. A. Segel, Rehovot

A selection of papers from recent issues:

Subscription Information:
To enter your subscription, or to request sample copies, contact Springer-Verlag, Dept. ZSW, Heidelberger Platz 3, D-1000 Berlin 33, W. Germany

Springer-Verlag
Berlin Heidelberg New York
London Paris Tokyo Hong Kong

Springer

Bio-mathematics

Managing Editor: S. A. Levin

Editorial Board: M. Arbib,
H. J. Bremermann, J. Cowan,
W. M. Hirsch, J. Karlin,
J. Keller, K. Krickeberg,
R. C. Lewontin, R. M. May,
J. D. Murray, A. Perelson,
T. Poggio, L. A. Segel

Springer-Verlag
Berlin Heidelberg New York
London Paris Tokyo Hong Kong

Springer